DETERGENCY OF SPECIALTY SURFACTANTS

SURFACTANT SCIENCE SERIES

96. Analysis of Surfactants: Second Edition, Revised and Expanded, *Thomas M. Schmitt*
97. Fluorinated Surfactants and Repellents: Second Edition, Revised and Expanded, *Erik Kissa*
98. Detergency of Specialty Surfactants, *edited by Floyd E. Friedli*
99. Physical Chemistry of Polyelectrolytes, *edited by Tsetska Radeva*

ADDITIONAL VOLUMES IN PREPARATION

Reactions and Synthesis in Surfactant Systems, *edited by John Texter*

Chemical Properties of Material Surfaces, *Marek Kosmulski*

Protein-Based Surfactants: Synthesis, Physicochemical Properties, and Applications, *edited by Ifendu A. Nnanna and Jiding Xia*

Oxide Surfaces, *edited by James A. Wingrave*

DETERGENCY OF SPECIALTY SURFACTANTS

edited by

Floyd E. Friedli

Goldschmidt Chemical Corporation
Dublin, Ohio

CRC Press
Taylor & Francis Group
Boca Raton London New York

CRC Press is an imprint of the
Taylor & Francis Group, an **informa** business

CRC Press
Taylor & Francis Group
6000 Broken Sound Parkway NW, Suite 300
Boca Raton, FL 33487-2742

First issued in paperback 2019

© 2001 by Taylor Francis Group, LLC
CRC Press is an imprint of Taylor & Francis Group, an Informa business

No claim to original U.S. Government works

ISBN-13: 978-0-8247-0491-9 (hbk)
ISBN-13: 978-0-367-39758-6 (pbk)

Visit the Taylor & Francis Web site at
http://www.taylorandfrancis.com

and the CRC Press Web site at
http://www.crcpress.com

Preface

This book summarizes the knowledge about the use of some nontraditional surfactants in detergents. The emphasis is on laundry detergents, with some discussion of other detergencies such as hard-surface cleaning and personal washing. The surfactants described in this volume are in the classic organic "head-tail" class of surfactants rather than polymers or silicone surfactants. The book originated from presentations given by noted scientists and authors at recent World Detergent Conferences in Montreux, Switzerland, and at AOCS and CESIO congresses.

Traditionally, powdered and liquid laundry detergents contained linear alkyl benzene sulfonates, ether sulfates, and alcohol ethoxylates as surfactants, along with builders, enzymes, polymers, and possibly bleaches as additional active ingredients. These formulations dealt well with different types of dirt and stains under a variety of water conditions.

In recent years, however, a number of specialty surfactants have been developed for use in detergents. Some of the materials are somewhat newer in a commercial sense, such as alkyl polyglucosides (APGs) and α-sulfomethyl esters. Other substances, such as alkyldiphenyl oxide disulfonates, betaines, alkanolamides, and ethoxylated amines, have been known for years and used routinely in other industries. In this volume, these ingredients are examined closely for advantages in detergents. Recent process improvements in producing new specialty surfactants such as ethoxylated methyl esters and alkyl ethylenediaminetriacetates have made their consideration as detergent ingredients necessary.

New specialty surfactants have been investigated and some adopted commercially for a variety of reasons—environmental, safety, performance, and formulation properties. In the environmental area, suppliers want surfactants that biodegrade rapidly and have low aquatic toxicity. The public also desires products with low human toxicity and minimal skin/eye irritation. There has always been a need for surfactants that clean well overall, clean certain stains well, or are cost effective. What criteria distinguish a commodity surfactant from a specialty one? In general, a specialty surfactant has these properties:

1. Smaller manufacturing volumes
2. Fewer producers
3. Special properties in some applications—low irritation, biodegradable, improved foam, etc.

Specialty surfactants are mostly secondary surfactants, not the primary ingredient in a formulation, and garner somewhat higher prices in the marketplace. Normally, over time a surfactant once considered a specialty slowly becomes a commodity. However, some surfactant classes can have both commodity and specialty uses. In fact, it is sometimes possible to convert a commodity into a specialty. For instance, linear alkylbenzene sulfonate (LAS) is considered a commodity. As such, it is sold either in the acid form at 96% active or as the sodium salt at 40% to 60% active. There are many producers of and applications for LAS. As a specialty, the sodium salt is sold as flakes at 90% active. Few producers and few applications need this material, but it commands a higher price. One main use is for dry blending in powdered detergents for smaller manufacturers, where it makes it possible to avoid purchase of expensive spray-drying equipment. Here the special property is its physical form.

The same is true of sodium lauryl sulfate (SLS). As a commodity, SLS is 30% active and has many producers and many applications. As a specialty, SLS is a powder that is 95% active and food grade. Again, there are few producers and few applications, the main one being toothpaste. Here the special property is its physical form and the fact that it is food grade.

In this volume, the special properties of specialty surfactants are their performance in detergents. Detergent formulations have changed immensely in the past 10 years. Both powders and liquids are more concentrated, making for better cleaning per gram. Enzyme use has increased in detergents, requiring surfactants that are compatible with enzymes or have a synergistic cleaning effect. Washing machines that use less water and only cold water are becoming more common in North America. Washing temperatures are decreasing dramatically in Europe also. These conditions require ingredients that are low foaming and that disperse easily to clean well in cold water.

In this volume of the Surfactant Series, originated by Martin Schick and published by Marcel Dekker, Inc., sugar-based surfactants such as alkyl polyglucosides are discussed in Chapter 1. Anionics such as alkyldiphenyl oxide disulfonates and α-sulfomethyl esters are presented in Chapters 3 and 4, and a number of new nonionics, amphoterics, chelating surfactants, and multifunctional ingredients in Chapters 2, 5, and 6. Special topics such as surfactants for dry cleaning, prespotters, and softeners are included in Chapters 7 through 9. The information contained herein should help the formulator choose the best ingredients for current detergent and cleaning formulas. This volume is related to the Surfactant Series' Volumes 59, on amphoterics, and 74 on novel surfactants, which can also be consulted.

Floyd E. Friedli

Contents

Contributors

Shoaib Arif   Home Care, Goldschmidt Chemical Corporation, Dublin, Ohio

Guy Broze   Personal Care, Colgate-Palmolive Research and Development, Inc., Milmort, Belgium

Régis César   Personal Care, Colgate-Palmolive Research and Development, Inc., Milmort, Belgium

Michael F. Cox   Research and Development, CONDEA Vista Company, Austin, Texas

Joe Crudden   Hampshire Chemical Corporation, Nashua, New Hampshire

Al Dabestani   Consultant, Dublin, Ohio

Amjad Farooq   Fabric Care–Advanced Technology, Colgate-Palmolive Company, Piscataway, New Jersey

Floyd E. Friedli   Research and Development Synthesis/Fabric Care, Goldschmidt Chemical Corporation, Dublin, Ohio

Terri Germain   Laundry and Cleaning Product Development, Stepan Company, Northfield, Illinois

T. J. Loughney   Samples LLC, Port Orchard, Washington

Jeffrey J. Mastrull   Fabric Care–Advanced Technology, Colgate-Palmolive Company, Piscataway, New Jersey

Ammanuel Mehreteab   Corporate Technology Center, Colgate-Palmolive Company, Piscataway, New Jersey

Lisa Quencer   Industrial Chemicals, The Dow Chemical Company, Midland, Michigan

Karl Schmid   Research and Development, Cognis GmbH, Düsseldorf, Germany

Holger Tesmann   Research and Development, Cognis GmbH, Düsseldorf, Germany

Michelle M. Watts   Research and Development Synthesis/Fabric Care, Goldschmidt Chemical Corporation, Dublin, Ohio

Upali Weerasooriya   Research and Development, CONDEA Vista Company, Austin, Texas

1
Alkyl Polyglycosides

KARL SCHMID and HOLGER TESMANN
Cognis GmbH, Düsseldorf, Germany

I. INTRODUCTION

A. History of Alkyl Polyglycosides

The first alkyl glucoside was synthesized and identified in the laboratory by Emil Fischer in 1893 [1]. This process is now well known as the "Fischer glycosidation" and comprises an acid-catalyzed reaction of glycoses with alcohols. The structure of ethyl glucoside was defined correctly by Fischer, as may be seen from the historical furanosidic formula proposed ("Fischer projection"). In fact, Fischer glycosidation products are complex, mostly equilibrium mixtures of α/β-anomers and pyranoside/furanoside isomers which also comprise randomly linked glycoside oligomers (Fig. 1) [2].

Accordingly, individual molecular species are not easy to isolate from Fischer reaction mixtures, which has been a serious problem in the past. After some improvement of this synthesis method [3], Fischer subsequently adopted the Koenigs-Knorr synthesis [4] for his investigations. Using this stereoselective glycosidation process introduced by W. Koenigs and E. Knorr in 1901, E. Fischer and B. Helferich in 1911 were the first to report the synthesis of a long-chain alkyl glucoside exhibiting surfactant properties [5]. As early as 1893, Fischer had noticed essential properties of alkyl glycosides, such as their high stability toward oxidation and hydrolysis, especially in strongly alkaline media. Both characteristics are valuable for alkyl polyglycosides in surfactant applications.

Research related to the glycosidation reaction is ongoing; some of the procedures for the synthesis of glycosides are summarized in Figure 2 [6]. The broad synthesis potential range has recently been reviewed in articles by Schmidt and Toshima and Tatsuta [7] and in a number of references cited there.

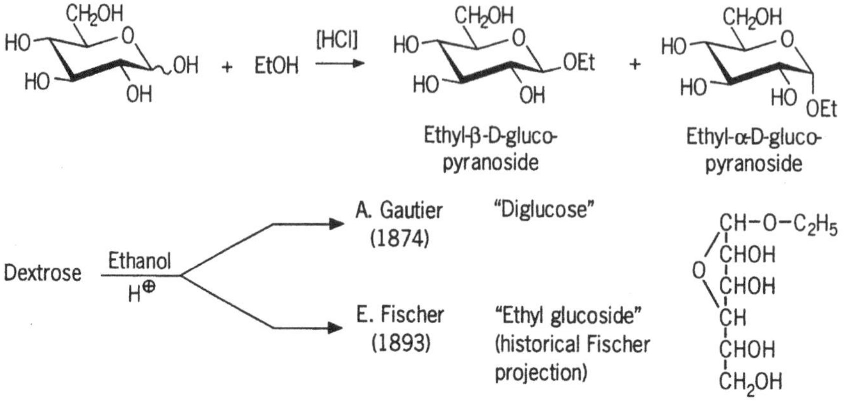

FIG. 1   Synthesis of glycosides according to Fischer.

B.   Developments in Industry

Alkyl glucosides—or alkyl polyglycosides, as the industrially manufactured products are widely known—are a classic example of products which, for a long time, were of academic interest only. The first patent application describing the use of alkyl glucosides in detergents was filed in Germany in 1934 [8]. Thereafter, another 40 to 50 years went by before research groups in various companies redirected their attention on alkyl glucosides and developed technical processes for the production of alkyl polyglycosides on the basis of the synthesis discovered by Fischer.

Rohm & Haas was the first to market an octyl/decyl polyglycoside in commercial quantities in the late seventies, followed by BASF and later SEPPIC. However, owing to the more hydrotropic character of this short-chain version as a surfactant, applications were limited to few market segments—for example, the industrial and institutional sectors.

The product quality of such short-chain alkyl polyglycosides has been improved in the past couple of years and new types of octyl/decyl polyglycoside are currently being offered by various companies, among them BASF, SEPPIC, Akzo Nobel, ICI, and Henkel.

At the beginning of the 1980s, several companies started programs to develop alkyl polyglycosides in a longer alkyl chain range (dodecyl/tetradecyl) with a view to making a new surfactant available to the cosmetics and detergent industries. They included Henkel KGaA, Düsseldorf, Germany, and Horizon, a division of A. E. Staley Manufacturing Company of Decatur, Illinois.

After the acquisition of Horizon by Henkel Corporation/USA a pilot plant went on line in 1988/1989 and was mainly intended to determine process para-

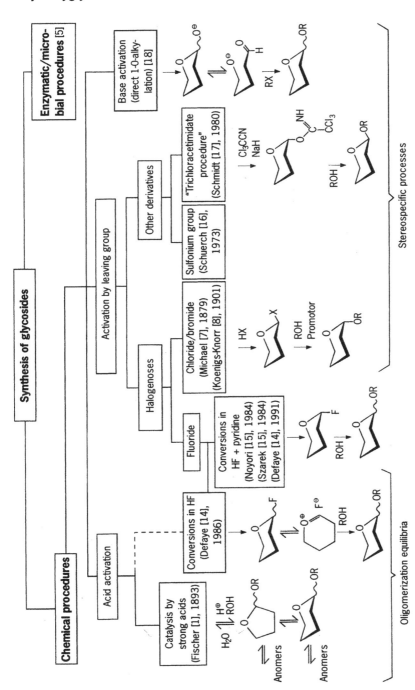

FIG. 2 Summary of methods for the synthesis of glycosides. (From Ref. 6.)

meters, to optimise product quality under industrial production conditions and to prepare the market for a new class of surfactants. New peaks in the commercial exploitation of alkyl polyglycosides (APG) were reached in 1992 with the inauguration of a 25,000 t.p.a. production plant for APG surfactants by Henkel Corporation in the United States and in 1995 with the opening of a second plant of equal capacity by Henkel KGaA in Germany [9].

II. TECHNOLOGY

Any production process suitable for use on an industrial scale must satisfy several criteria. The ability to produce products with suitable performance properties and process economy are the most important. There are other aspects, such as minimizing side reactions or waste and emissions. The technology used should have a flexibility which allows product properties and quality features to be adapted to market requirements.

So far as the industrial production of alkyl polyglycosides is concerned, processes based on the Fischer synthesis have been successfully adopted. Development work over 20 years has enabled the efficiency of this synthesis route to be increased to a level where it has finally become attractive for industrial application. Optimization work, particularly in the use of long-chain alcohols, such as dodecanol/tetradecanol, has resulted in distinct improvements in product quality and process economy.

Modern production plants built on the basis of the Fischer synthesis are the embodiment of low-waste, virtually emission-free technologies. Another advantage of the Fischer synthesis is that the average degree of polymerization of the products can be precisely controlled over a wide range. Relevant performance properties, for example hydrophilicity or water solubility, can thus be adapted to meet applicational requirements. Additionally the raw material base is no longer confined to water-free glucose [10–12].

A. Raw Materials for the Manufacture of Alkyl Polyglycosides

1. Fatty Alcohols

Fatty alcohols can be obtained either from petrochemical sources (synthetic fatty alcohols) or from natural, renewable resources, such as fats and oils (natural fatty alcohols). Fatty-alcohol blends are used in the alkyl polyglycoside synthesis to build up the hydrophobic part of the molecule. The natural fatty alcohols are obtained after transesterification and fractionation of fats and oils (triglycerides), leading to the corresponding fatty acid methyl esters, and subsequent hydrogenation. Depending on the desired alkyl chain length of the fatty alcohol, the main feedstocks are oils and fats of the following composition: coconut or palm kernel oil for the $C_{12/14}$ range and tallow, palm, or rapeseed oil for the $C_{16/18}$ fatty alcohols.

2. Carbohydrate Source

The hydrophilic part of the alkyl polyglycoside molecule is derived from a carbohydrate. Based on starch from corn, wheat, or potatoes, both polymeric and monomeric carbohydrates are suitable as raw materials for the production of alkyl polyglycosides. Polymeric carbohydrates include, for example, starch or glucose syrups with low degradation levels while monomeric carbohydrates can be any of the various forms in which glucose is available, for example water-free glucose, glucose monohydrate (dextrose), or highly degraded glucose syrup. Raw material choice influences not only raw material costs, but also production costs. Generally speaking, raw material costs increase in the order starch/glucose, syrup/glucose, and monohydrate/water-free glucose whereas plant equipment requirements and hence production costs decrease in the same order (Fig. 3).

3. Degree of Polymerization

Through the polyfunctionality of the carbohydrate partner, the conditions of the acid-catalyzed Fischer reaction yield an oligomer mixture in which on average more than one glycose unit is attached to an alcohol molecule. The average number of glycose units linked to an alcohol group is described as the degree of poly-

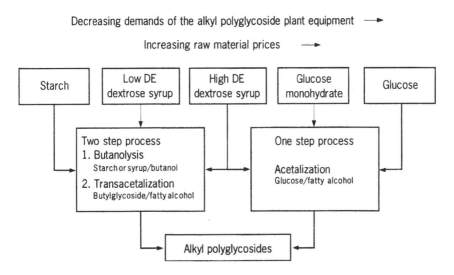

FIG. 3 Carbohydrate sources for industrial-scale alkyl polyglycoside synthesis. DE = dextrose equivalent.

merization (DP). Figure 4 shows the distribution for an alkyl polyglycoside with DP = 1.3. In this mixture, the concentration of the individual oligomers (mono-, di-, tri-, . . .-, glycoside) is largely dependent on the ratio of glucose to alcohol in the reaction mixture. The average DP is an important characteristic with regard to the physical chemistry and application of alkyl polyglycosides. In an equilibrium distribution, the DP—for a given alkyl chain length—correlates well with basic product properties, such as polarity, solubility, etc. In principle, this oligomer distribution can be described by a mathematical model. P. M. McCurry [13] showed that a model developed by P. J. Flory [14] for describing the oligomer distribution of products based on polyfunctional monomers can also be applied to alkyl polyglycosides. This modified version of the Flory distribution describes alkyl polyglycosides as a mixture of statistically distributed oligomers.

The content of individual species in the oligomer mixture decreases with increasing degree of polymerization. The oligomer distribution obtained by this mathematical model accords well with analytical results.

B. Production of Alkyl Polyglycosides

Basically, all processes for the reaction of carbohydrates to alkyl polyglycosides by the Fischer synthesis can be attributed to two process variants, namely direct

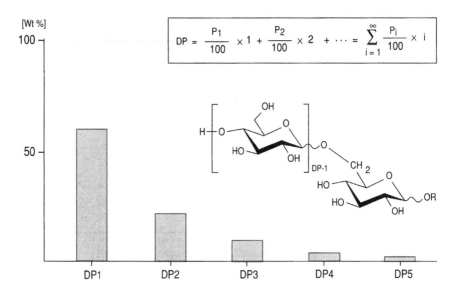

$$DP = \frac{P_1}{100} \times 1 + \frac{P_2}{100} \times 2 + \cdots = \sum_{i=1}^{\infty} \frac{P_i}{100} \times i$$

FIG. 4 Typical distribution of dodecyl glycoside oligomers in a DP = 1.3 mixture. R = dodecyl.

synthesis and the transacetalization process. In either case, the reaction can be carried out in batches or continuously.

Direct synthesis is simpler from the equipment point of view [15–17]. In this case, the carbohydrate reacts directly with the fatty alcohol to form the required long-chain alkyl polyglycoside. The carbohydrate used is often dried before the actual reaction (for example, to remove the crystal-water in case of glucose monohydrate = dextrose). This drying step minimizes side reactions which take place in the presence of water.

In the direct synthesis, monomeric solid glucose types are used as fine-particle solids. Since the reaction is a heterogeneous solid/liquid reaction, the solid has to be thoroughly suspended in the alcohol.

Highly degraded glucose syrup (DE>96; DE=dextrose equivalents) can react in a modified direct synthesis. The use of a second solvent and/or emulsifiers (for example, alkyl polyglycoside) provides for a stable fine-droplet dispersion between alcohol and glucose syrup [18,19].

The two-stage transacetalization process involves more equipment than the direct synthesis. In the first stage, the carbohydrate reacts with a short-chain alcohol (for example n-butanol or propylene glycol) and optionally depolymerizes. In the second stage, the short-chain alkyl glycoside is transacetalized with a relatively long-chain alcohol ($C_{12/14}$-OH) to form the required alkyl polyglycoside. If the molar ratios of carbohydrate to alcohol are identical, the oligomer distribution obtained in the transacetalization process is basically the same as in the direct synthesis.

The transacetalization process is applied if oligo- and polyglycoses (for example, starch, syrups with a low DE value) are used [20]. The necessary depolymerization of these starting materials requires temperatures of > 140°C. Depending on the alcohol used, this can create higher pressures which impose more stringent demands on equipment and can lead to a higher plant cost.

Generally for the same capacity, the transacetalization process results in higher plant costs than the direct synthesis. Besides the two reaction stages, additional storage tanks and recycle facilities for the short-chain alcohol are needed. Alkyl polyglycosides have to be subjected to additional or more elaborate refining on account of specific impurities in the starch (for example, proteins). In a simplified transacetalization process, syrups with a high glucose content (DE>96 %) or solid glucose types can react with short-chain alcohols under normal pressure [21–25]. Continuous processes have been developed on this basis [23]. Figure 5 shows both synthesis routes for alkyl polyglycosides.

The requirements for the design of alkyl polyglycoside production plants based on the Fischer synthesis are critically determined by the carbohydrate types used and by the chain length of the alcohol used as described here for the production of alkyl polyglycosides on the basis of octanol/decanol and dodecanol/tetradecanol.

Under the conditions of the acid-catalyzed syntheses of alkyl polyglycosides, secondary products, such as polydextrose [26,27], ethers, and colored impurities,

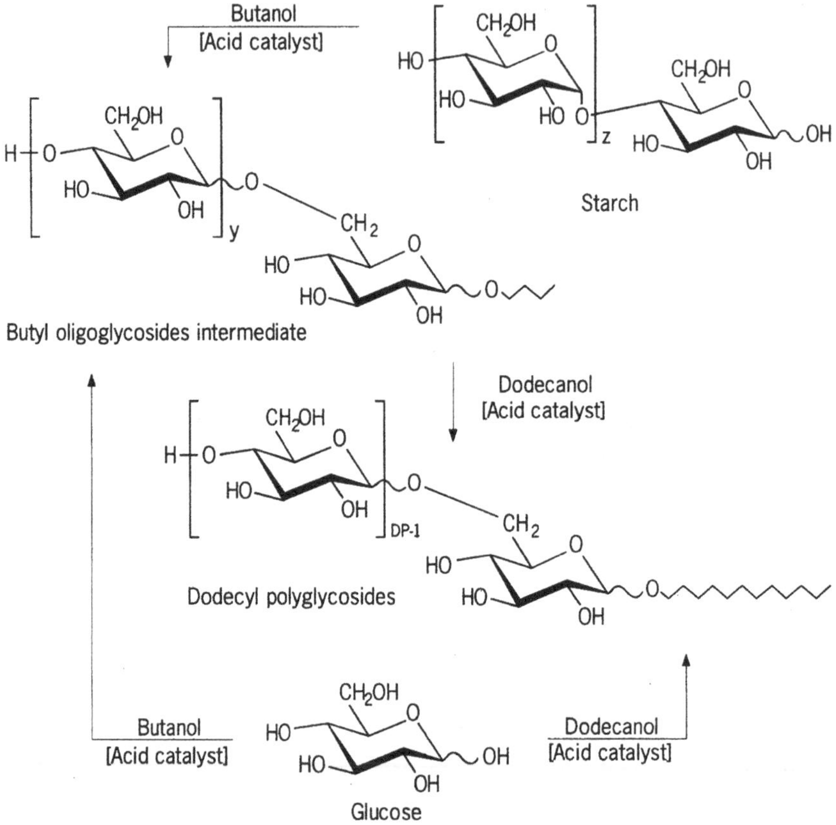

FIG. 5 Alkyl polyglycoside surfactants—industrial synthesis pathways.

are formed. Polydextroses are substances of undefined structure which are formed in the course of the synthesis through the polymerization of glycoses. The type and concentration of the substances formed by secondary reactions are dependent on process parameters, such as temperature, pressure, reaction time, catalyst, etc. One of the problems addressed by development work on industrial alkyl polyglycoside production over recent years was to minimize this synthesis-related formation of secondary products.

Generally, the production of alkyl polyglycosides based on short-chain alcohols ($C_{8/10}$-OH) and with a low DP (large alcohol excess) presents the fewest problems. Fewer secondary products are formed with the increasing excess of alcohol in the reaction stage. The thermal stress and formation of pyrolysis products during removal of the excess alcohol are reduced.

The Fischer glycosidation may be described as a process in which, in a first step, the dextrose reacts relatively quickly and an oligomer equilibrium is reached. This step is followed by slow degradation of the alkyl polyglycoside. In the course of the degradation, which consists of dealkylation and polymerization steps, the thermodynamically more stable polydextrose is formed substantially irreversibly in increasing concentrations. Reaction mixtures which have exceeded an optimal reaction time may be described as "overreacted."

If the reaction is terminated too early, the resulting reaction mixture contains a significant amount of residual dextrose. The loss of alkyl polyglycoside active substance in the reaction mixture correlates well with the formation of polydextrose, the reaction mixture in the case of overreacted systems gradually becoming heterogeneous again through precipitating polydextrose. Accordingly, product quality and product yield are critically influenced by the time at which the reaction is terminated. Starting with solid dextrose, alkyl polyglycosides low in secondary products are obtained, providing other polar constituents (polydextrose) are filtered off together with the remaining carbohydrate from a reaction mixture that has not fully reacted [28,29].

In an optimized process, the concentration of secondary products formed by etherification remains relatively low (depending on reaction temperature and time, type and concentration of catalyst, etc.). Figure 6 shows the typical course of a direct reaction of dextrose and fatty alcohol ($C_{12/14}$-OH).

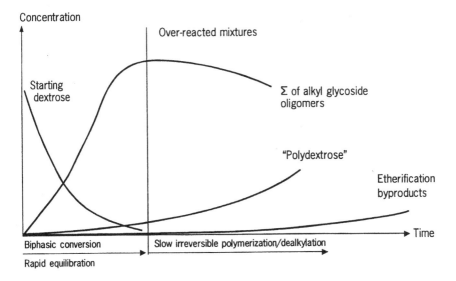

FIG. 6 Mass balance of the glycosidation process.

In the Fischer glycosidation, the reaction parameters temperature and pressure are closely related. To produce an alkyl polyglycoside low in secondary products, pressure and temperature have to be adapted to one another and carefully controlled. Low reaction temperatures (<100°C) in the acetalization lead to alkyl polyglycosides low in secondary products. However, low temperatures result in relatively long reaction times (depending on the chain length of the alcohol) and low specific reactor efficiencies.

Relatively high reaction temperatures (>100°C, typically 110–120°C) can lead to changes in color of the carbohydrates. By removing the lower-boiling reaction products (water in the direct synthesis, short-chain alcohols in the transacetalization process) from the reaction mixture, the acetalization equilibrium is shifted to the product side. If a relatively large amount of water is produced per unit of time, for example by high reaction temperatures, provision has to be made for the effective removal of this water from the reaction mixture. This minimizes secondary reactions (particularly the formation of polydextrose) which take place in the presence of water. The evaporation efficiency of a reaction stage depends not only on pressure, but also on temperature and on the design of the reactor (stirrer, heat-exchange area, evaporation area, etc.). Typical reaction pressures in the transacetalization and direct synthesis variants are between 20 and 100 mbar.

Another important optimization factor is the development of selective catalysts for the glycosidation process so that for example the formation of polydextrose and etherification reactions can be suppressed. As already mentioned, acetalization or transacetalization in the Fischer synthesis is catalyzed by acids. In principle, any acids with sufficient strength are suitable for this purpose, such as sulfuric acid, paratoluene- and alkylbenzene sulfonic acid, and sulfosuccinic acid. The reaction rate depends on the acidity and the concentration of the acid in the alcohol.

After the reaction, the acidic catalyst is neutralized by a suitable base, for example sodium hydroxide or magnesium oxide. The neutralized reaction mixture is a yellowish solution containing 50 to 80 % fatty alcohol. The high fatty-alcohol content results from the molar ratios of carbohydrate to fatty alcohol. This ratio is adjusted to obtain a specific DP for the technical alkyl polyglycosides and is generally between 1:2 and 1:6. The excess fatty alcohol is removed by vacuum distillation. Important boundary conditions include:

1.   Residual fatty alcohol content in the product must be <1% because otherwise solubility and odor are adversely affected.

2.   To minimize the formation of unwanted pyrolysis products or discoloring components, thermal stressing and residence time of the target product must be kept as low as possible in dependence upon the chain length of the alcohol.

3.   No monoglycoside should enter the distillate because the distillate is recycled in the reaction as pure fatty alcohol.

In case of dodecanol/tetradecanol these requirements for the removal of excess fatty alcohol are largely satisfied by multistage distillation using thin-layer or short-path evaporators [30,31]. In these evaporators, the mechanically moved film provides for high specific evaporation efficiency and a short product residence time and at the same time a good vacuum. The end product after distillation is an almost pure alkyl polyglycoside which accumulates as a solid with a melting range of 70–150°C. Figure 7 summarizes the main process steps for the synthesis of alkyl polyglycosides.

After removal of the fatty alcohol, the alkyl polyglycoside active substance is directly dissolved in water so that a highly viscous 50% to 70% alkyl polyglycoside paste is formed. In subsequent refining steps, this paste is worked up into a product of satisfactory quality in accordance with performance-related requirements. These refining steps may comprise bleaching of the product, the adjustment of product characteristics, such as pH value and active substance content, and microbial stabilization. The effort and hence the cost involved in these process steps to obtain certain quality features, such as color, depend on performance requirements, on the starting materials, the DP required, and the quality of the process steps. Figure 8 illustrates an industrial production process for long-chain alkyl polyglycosides ($C_{12/14}$) via direct synthesis.

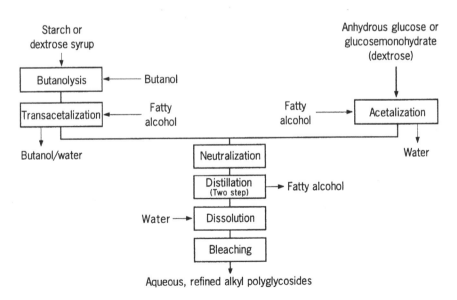

FIG. 7 Simplified flow diagram for the production of alkyl polyglycosides based on different carbohydrate sources—direct synthesis and transacetalization.

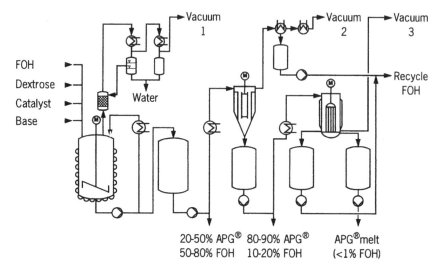

FIG. 8 Typical industrial-scale glycosidation process for $C_{12/14}$-APG. FOH = fatty alcohol.

C. Analytical Characterization of Alkyl Polyglycosides

Analytical methods for alkyl polyglycosides were published by H. Waldhoff, J. Scherler, and M. Schmitt [32] and by other groups in 1995 and 1996 [33–36]. Commercial alkyl polyglycosides are complex mixtures of species which differ mainly in the degree of polymerization (a typical distribution is given in Table 1) and in the length of the alkyl chains.

Alkyl monoglycosides are the main group of components with a content of > 50%, followed by the diglycosides and higher oligomers up to heptaglycosides. Small amounts of more highly glycosidated species are also present. Species with a degree of glycosidation > 5 are not normally determined in routine analysis because the amounts involved are too small. The most important analytical techniques routinely used for the characterization of main and trace components in commercial alkyl polyglycosides are high-performance liquid chromatography (HPLC) and gas chromatography (GC).

1. Alkyl Polyglycoside Determination by GC

The GC technique which has proven to be particularly suitable for the analysis of alkyl mono- and oligoglycosides is high-temperature gas chromatography (HTGC). HTGC uses temperatures of up to 400°C, which enables oligomeric alkyl polyglycosides up to the very high boiling heptaglycosides to be analyzed. The hydroxyl groups in alkyl polyglycosides have to be converted into silyl

TABLE 1 Composition of Two Alkyl
Polyglycoside Surfactants

Substances	Sample 1	Sample 2
Monoglycosides	65%	51%
Diglycosides	19%	19%
Triglycosides	9%	13%
Tetraglycosides	5%	10%
Pentaglycosides	2%	4%
Hexaglycosides	0% (<0.5%)	2%
Heptaglycosides	0% (<0.5%)	1%

ethers before analysis to prevent sample decomposition. A typical HTGC of a commercial alkyl polyglycoside sample is shown in Figure 9.

2.   Alkyl Polyglycoside Characterization by HPLC

Alkyl polyglycoside characterization by HPLC is routinely performed using an isocratic reversed-phase system. In most cases, no particular sample preparation is necessary; after dissolution in the eluent, the sample solution is filtered and directly injected into the system. A typical chromatogram and the chromatographic conditions are shown in Figure 10.

The retention time corresponds to the lipophilicity of the substances separated. The individual species are identified and quantified by the external standard method using commercially available alkyl glycosides for calibration. The alkyl monoglycosides are separated cleanly enough to allow sufficiently accurate quantification. Detailed determination of individual oligoglycosides and separation into α and β anomers, pyranosides, and furanosides are only possible with a more polar mobile phase that requires tediously long analysis times [37]. The analysis of alkyl glycosides in commercial alkyl polyglycoside products by HPLC provides good results for analytical tasks which do not require high resolution of a broad spectrum of components. Typical applications include raw material identification, comparative alkyl polyglycoside analysis, quantifications, and calculations solely on the basis of the alkyl monoglycoside contents.

III.   PHYSICOCHEMICAL PROPERTIES OF ALKYL POLYGLYCOSIDES

A.   Surface Tension

The surface tension of alkyl (poly)glycosides was investigated as a function of the alkyl chain and the degree of polymerization (DP) using samples differing in composition. In addition to pure surfactants for characterizing the basic depen-

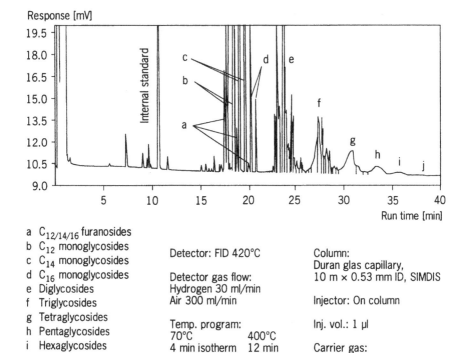

a $C_{12/14/16}$ furanosides
b C_{12} monoglycosides
c C_{14} monoglycosides
d C_{16} monoglycosides
e Diglycosides
f Triglycosides
g Tetraglycosides
h Pentaglycosides
i Hexaglycosides
j Heptaglycosides

Detector: FID 420°C

Detector gas flow:
Hydrogen 30 ml/min
Air 300 ml/min

Temp. program:
70°C   400°C
4 min isotherm   12 min
10°C/min   isotherm

Column:
Duran glas capillary,
10 m × 0.53 mm ID, SIMDIS

Injector: On column

Inj. vol.: 1 µl

Carrier gas:
Hydrogen 20 ml/min

FIG. 9   HTGC chromatogram of long-chain alkyl polyglycoside ($C_{12/14}$-APG).

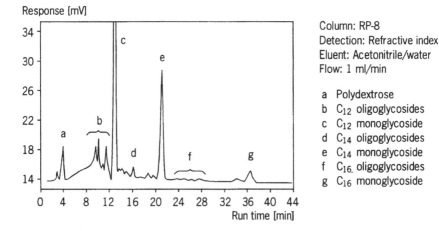

Column: RP-8
Detection: Refractive index
Eluent: Acetonitrile/water
Flow: 1 ml/min

a Polydextrose
b C_{12} oligoglycosides
c C_{12} monoglycoside
d C_{14} oligoglycosides
e C_{14} monoglycoside
f C_{16} oligoglycosides
g C_{16} monoglycoside

FIG. 10   HPLC chromatogram of $C_{12/14}$-APG sample.

dencies, numerous data collections on surfactants of technical quality are available. Shinoda et al. [38] and Böcker et al. [39] determined the critical micelle concentration (CMC) and the surface tension values at the CMC from surface tension/concentration curves. A few selected values are set out in Table 2.

Figure 11 shows the surface tension as a function of concentration for three alkyl monoglycosides (C_nG_1) and a technical $C_{12/14}$-APG at 60°C [40]. The CMC values of the pure alkyl polyglycosides and the technical alkyl polyglycosides are comparable with those of typical nonionic surfactants and decrease distinctly with increasing alkyl chain length. As the figures set out in Table 2 show, the alkyl chain length has a far stronger influence on the CMC by comparison with the number of glucoside groups of the alkyl polyglycosides.

The surface tension behavior of mixtures of alkyl polyglycosides and anionic surfactants was investigated with reference to an alkyl polyglycoside/fatty-alcohol sulfate (FAS) mixture. Figure 12 shows the surface tension/concentration curves at 60°C for $C_{12/14}$-APG, $C_{12/14}$-FAS, and two mixtures of these

TABLE 2 cmc Values from Surface Tension
Measurements

Substance	Temperature (°C)	cmc (mol/L)
β-D-C_8G_1	25	$2.5 \cdot 10^{-2}$ [15]
C_8G_1	25	$1.8 \cdot 10^{-2}$ [16]
β-D-$C_{10}G_1$	25	$2.2 \cdot 10^{-3}$ [15]
β-D-$C_{12}G_1$	25	$1.9 \cdot 10^{-4}$ [15]
$C_{12}G_1$	60	$1.7 \cdot 10^{-4}$ [16]

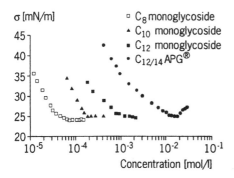

FIG. 11 Static surface tension of alkyl glycosides with different alkyl chain lengths as a function of the concentration in distilled water at 60°C.

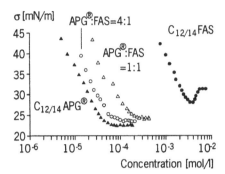

FIG. 12 Static surface tension of $C_{12/14}$-APG and $C_{12/14}$ FAS and 1:1 and 4:1 mixtures thereof as a function of the concentration in distilled water at 60°C. (From Ref. 17.)

surfactants, as measured at 60°C [40]. The values of the mixtures are close to the curve for alkyl polyglycoside even despite a high anionic surfactant content. This corresponds to the normally observed behavior of mixtures of anionic and nonionic surfactants differing considerably in their CMC values [41].

B. Phase Behavior

The nonionic class of alkyl polyglycosides differs from fatty-alcohol ethoxylates by its characteristic structure which considerably affects the association of molecules in solution, the phase behavior, and the interfacial activity. The hydrophilic head of the glucose ring, associated with the surrounding water molecules by hydrogen bonds, is rather voluminous as compared to the alkyl chain. However, the hydration is low compared to fatty alcohol ethoxylates. Therefore, basic phenomena like cloud point, thermal phase inversion, or gel formation in medium concentrations of pure solutions cannot be observed in case of the alkyl polyglycosides.

The phase behavior of alkyl polyglycosides in aqueous solution gives important hints for the handling and formulation of a product in regard to viscosity, flow behavior, and phase stability. The phase stability was investigated in pure solutions and systems mixed with alkyl ether sulfate, alkyl sulfate, and alkyl sulfate/fatty alcohol. The phase diagram of $C_{12/14}$-APG shows the main important characteristics (Fig. 13). The Krafft temperature of a fresh solution of $C_{12/14}$-APG in water is approximately 20–22°C at low concentration and increases to ~ 37°C during storage time. Below the Krafft temperature a highly viscous dispersion of crystals and brine exists, whereas above a clear micellar solution exists (L_1) in a wide range of concentration without lamellar phases. The consistency of the L_1 phase is like honey and shows Newtonian flow behavior.

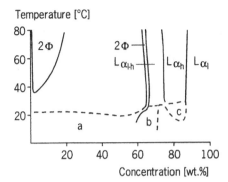

FIG. 13   Phase diagram of the $C_{12/14}$-APG/water system.

At low concentration the micellar solution is limited by a two-phase area $(w+L_1)$ where a surfactant rich and a water rich phase coexist (coacervate). The area of coacervates is enlarged toward lower temperature and higher concentration if electrolytes are added. It is destroyed by addition of anionics. Similarly, the Krafft temperature decreases in the presence of ionic or other nonionic surfactants, so that neither phase separation by coacervate formation nor crystallization is observed at concentrations typically used in cleansing formulations.

C. Microemulsion Phases

The solubilization of comparably small amounts of oil components in rinse and shampoo formulations demonstrates the basic emulsification properties which alkyl polyglycosides should be expected to show as nonionic surfactants. However, a proper understanding of phase behavior in multicomponent systems is necessary in order to evaluate alkyl polyglycosides as powerful emulsifiers in combination with suitable hydrophobic coemulsifiers [42].

In general, the interfacial activity of alkyl polyglycosides is determined by the carbon chain length and, to a lesser extent, by the degree of polymerization (DP). Interfacial activity increases with the alkyl chain length and is at its highest near or above the CMC with a value below 1 mN/m. At the water/mineral oil interface, $C_{12/14}$-APG shows lower surface tension than $C_{12/14}$ alkyl sulfate [43]. Interfacial tensions of n-decane, isopropyl myristate, and 2-octyl dodecanol have been measured for pure alkyl monoglucosides (C_8, C_{10}, C_{12}), and their dependence on the solubility of alkyl polyglycosides in the oil phase has been described [44]. Medium-chain alkyl polyglycosides may be used as emulsifiers for o/w emulsions in combination with hydrophobic coemulsifiers [45].

Alkyl polyglycosides differ from ethoxylated nonionic surfactants in that they do not undergo temperature-induced phase inversion from oil-in-water (o/w) to water-in-oil (w/o) emulsions. Instead, their hydrophilic/lipophilic properties can be balanced by mixing with a hydrophobic emulsifier, such as glycerol monooleate (GMO) or sorbitan monolaurate (SML).

In fact, the phase behavior and interfacial tension of such alkyl polyglycoside emulsifier systems closely resemble those of conventional fatty alcohol ethoxylate systems if temperature as a key phase behavior parameter is replaced by the mixing ratio of the hydrophilic/lipophilic emulsifiers in the non-ethoxylated system [46].

The system of dodecane, water, and lauryl glucoside with sorbitan laurate as a hydrophobic coemulsifier forms microemulsions at a certain mixing ratio of $C_{12/14}$-APG to SML of 4:6 to 6:4 (Fig. 14). Higher SML contents lead to w/o emulsions whereas higher alkyl polyglycoside contents produce o/w emulsions. Variation of the total emulsifier concentration results in a so-called Kahlweit fish in the phase diagram, the body containing three-phase microemulsions and the tail single-phase microemulsions, as observed with ethoxylated emulsifiers as a function of temperature. The high emulsifying capacity of the $C_{12/14}$-APG/SML mixture as compared with a fatty alcohol ethoxylate system is reflected in the fact that even 10% of the emulsifier mixture is sufficient to form a single-phase microemulsion.

The similarity in the phase inversion pattern of both surfactant types is not limited to phase behavior, but can also be found in the interfacial tension of the emulsifying systems. The hydrophilic-lipophilic properties of the emulsifier mixture are balanced at an $C_{12/14}$-APG/SML ratio of 4:6 where interfacial tension

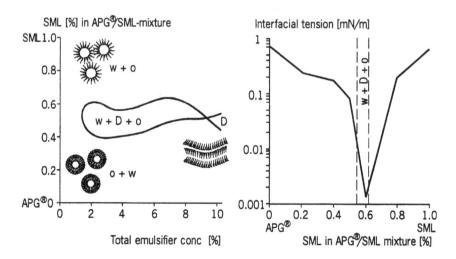

FIG. 14 Phase behavior and interfacial tension of a Lauryl Glucoside ($C_{12/14}$-APG)/Sorbitan Laurate/Dodecane mixture in water ($C_{12}H_{26}$:water = 1:1 at 25°C).

is at its lowest. It is remarkable that a very low minimum interfacial tension (around 10^{-3} mN/m) is observed with the $C_{12/14}$-APG/SML mixture.

In the case of microemulsions containing alkyl polyglycosides, the high interfacial activity is attributable to the fact that the hydrophilic alkyl polyglycoside with its large glucoside head group is mixed in the ideal ratio with a hydrophobic coemulsifier having a smaller head group at the oil-water interface. In contrast to ethoxylated nonionic surfactants, hydration (and hence the effective size of the head group) is not noticeably temperature dependent. Accordingly, paralleling interfacial tension, only a slight dependence on temperature is observed for the phase behavior of the nonethoxylated emulsifier mixture [46].

This provides for interesting new formulation systems because, in contrast to fatty-alcohol ethoxylates, temperature-stable microemulsions can be formed with alkyl polyglycosides. By varying the surfactant content, the type of surfactant used, and the oil/water ratio, microemulsions can be produced with customized performance properties, such as transparency, viscosity, refatting effect, and foaming behavior. In mixed systems of alkyl ether sulfates and nonionic coemulsifiers, extended microemulsion areas are observed and may be used for the formulation of concentrates or fine-particle o/w emulsions [42,46].

An evaluation has been made of pseudoternary phase triangles of multicomponent systems containing alkyl polyglycoside/SLES and SML with a hydrocarbon (Dioctyl Cyclohexane*) [42] and alkyl polyglycoside/SLES and GMO with polar oils (Dicaprylyl Ether/Octyl Dodecanol) [46]. They demonstrate the variability and extent of areas for o/w, w/o, or microemulsions for hexagonal phases and for lamellar phases in dependence on the chemical structure and mixing ratio of the components.

If these phase triangles are superimposed on congruent performance triangles, indicating, for example, foaming behavior and viscosity properties of the corresponding mixtures, they provide a valuable aid for the formulator in finding specific and well-designed microemulsion formulations for facial cleansers or refatting foam baths, for example. As an example, a suitable microemulsion formulation for refatting foam baths can be derived from the phase triangle in Figure 15. The oil mixture consists of Dicaprylyl Ether and Octyl Dodecanol in a ratio of 3:1. The hydrophilic emulsifier is a 5:3 mixture of Coco Glucoside (C_{8-14}-APG) and sodium laureth sulfate (SLES). This high-foaming surfactant mixture forms the basis of body-cleansing formulations. The hydrophobic coemulsifier is Glyceryl Oleate (GMO). The water content is kept constant at 60%.

Starting from an oil- and coemulsifier-free system, a 40% C_{8-14}-APG/SLES mixture in water forms a hexagonal liquid crystal (H_{Ia}). The surfactant paste is highly viscous and nonpumpable at 25°C.

---

*Capital letters are used for INCI names throughout chapter

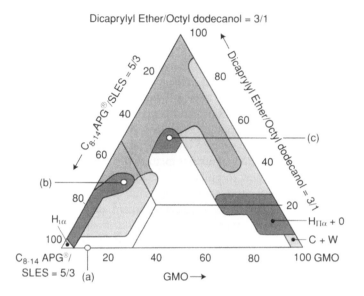

FIG. 15 Pseudoternary phase triangle of a six-component system (60% water, 25°C).

Only a fraction of the C_{8-14}-APG/SLES mixture needs to be replaced by the hydrophobic cosurfactant GMO to obtain a lamellar phase of medium viscosity (L_a, point (a) with a value of 23,000 mPa·s at 1 s⁻¹). In terms of practical application, this means that the high-viscosity surfactant paste changes into a pumpable surfactant concentrate. Despite the increased GMO content, the lamellar phase remains intact.

However, the viscosity increases significantly and reaches levels for the liquid gel which are even above those of the hexagonal phase. In the GMO corner, the mixture of GMO and water forms a solid cubic gel. When oil is added, an inverse hexagonal liquid (H_{IIa}) is formed with water as the internal phase.

The hexagonal liquid crystal rich in surfactants and the lamellar liquid crystal differ considerably in their reactions to the addition of oil. Whereas the hexagonal liquid crystal can only take up very small quantities of oil, the lamellar phase area extends far towards the oil corner. The capacity of the lamellar liquid crystal to take up oil clearly increases with increasing GMO content.

Microemulsions are only formed in systems with low GMO contents. An area of low-viscosity o/w microemulsions extends from the C_{8-14}-APG/SLES corner along the surfactant/oil axis up to an oil content of 14%. At point (b), the microemulsion consists of 24% surfactants, 4% coemulsifier, and 12% oil, representing an oil-containing surfactant concentrate with a viscosity of 1600 mPa·s at 1 s⁻¹.

The lamellar area is followed by a second microemulsion are at point (c). This microemulsion is an oil-rich gel with a viscosity of 20,000 mPa·s at 1 s^{-1} (12% surfactants, 8% coemulsifier, 20% oils) and is suitable as a refatting foam bath. The C$_{8-14}$-APG/SLES mixture contributes toward (cleansing performance and foam while the oil mixture acts as a refatting skin care component.

In order to obtain a refatting effect with a microemulsion, the oil must be released; i.e., the microemulsion must break up during application. A microemulsion of suitable composition breaks up during rinsing when it is heavily diluted with water, thus releasing the oil for refatting effects on the skin.

In conclusion, it may be said that microemulsions can be produced with alkyl polyglycosides in combination with suitable coemulsifiers and oil mixtures. They are distinguished by their transparency and by their high temperature stability, high storage stability, and high solubilizing capacity for oils.

The properties of alkyl polyglycosides with comparably long alkyl chains (C$_{16}$ to C$_{22}$) as o/w emulsifiers are even more pronounced. In conventional emulsions with fatty alcohol or glyceryl stearate as coemulsifier and consistency regulator, long-chain alkyl polyglycosides show better stability than the medium-chain C$_{12/14}$-APG described above. Technically, the direct glycosidation of C$_{16/18}$ fatty alcohol leads to a mixture of C$_{16/18}$ alkyl polyglycoside and cetearyl alcohol from which cetearyl alcohol cannot be completely distilled off by usual techniques to avoid color and odor deterioration. Utilizing the residual cetearyl alcohol as co-emulsifier, self-emulsifying o/w bases containing 20% to 60% C$_{16/18}$ alkyl polyglycoside are the most suitable in practice for formulating cosmetic cremes and lotions based entirely on vegetable raw materials. Viscosity is easy to adjust through the amount of C$_{16/18}$ alkyl polyglycoside/cetearyl alcohol compound and excellent stability is observed, even in the case of highly polar emollients, such as triglycerides [47].

IV. ALKYL POLYGLYCOSIDES IN HARD-SURFACE CLEANERS AND LAUNDRY DETERGENTS

A. Alkyl Polyglycosides in Manual Dishwashing Detergents

Since the introduction of manual dishwashing detergents (MDD), the requirements the consumer expects of this group of products have continually changed. For modern MDDs the consumer wants different aspects to be considered according to his/her personal relevance (Fig. 16). The possibility of using alkyl polyglycosides on a commercial scale began with the development of economic production processes and the establishment of large capacity production plants.

Alkyl polyglycosides with an alkyl chain length of C$_{12/14}$ is preferred for MDDs. The typical average DP is around 1.4. For the product developer, alkyl polyglycosides have a number of interesting properties [48–52]:

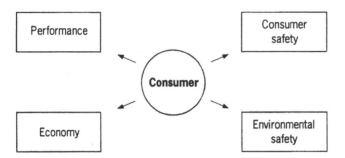

FIG. 16   Requirements for manual dishwashing detergents.

Synergistic performance interactions with anionic surfactants
Good foaming behavior
Low skin irritation potential
Excellent ecological and toxicological properties
Completely derived from renewable resources

The basic properties of alkyl polyglycosides in combination with other surfactants are discussed and their use in the individual markets of Europe, North and Latin America, and Asia is described in the upcoming sections.

1.   Basic Properties of Alkyl Polyglycosides

(a)   Dishwashing Performance.   In conjunction with anionic surfactants, alkyl polyglycosides show significant synergistic effects which can be demonstrated not only by physicochemical methods, but also by methods of greater relevance to the consumer, for example, plate test. Typical soils in the plate test are fats (as sole soil component) and so-called mixed soils (mixtures of fat, starch and protein). $C_{12/14}$-APG show pronounced synergisms with the three primary surfactants—linear alkyl benzene sulfonate (LAS), secondary alkane sulfonate (SAS), and fatty-alcohol sulfate (FAS). These synergisms are far more pronounced than those observed with fatty-alcohol ether sulfate (FAES) (Fig. 17). In contrast to alkyl polyglycosides, other nonionic surfactants, such as fatty-alcohol polyethylene glycol ether (FAEO), do not show any synergisms with FAES (Fig. 18).

Three primary surfactant systems—LAS, SAS, and FAS—are found worldwide in manual dishwashing detergents, often in combination with FAES. The performance of these primary surfactant systems can be enhanced by so-called cosurfactants, for example, betaines or fatty-acid alkanolamides and, especially, alkyl polyglycosides (Fig. 19) [51].

The dishwashing performance of the FAS/FAES-based surfactant system, which is relatively poor in contrast to the LAS/FAES and SAS/FAES systems, can be significantly increased by replacing small quantities of the primary sur-

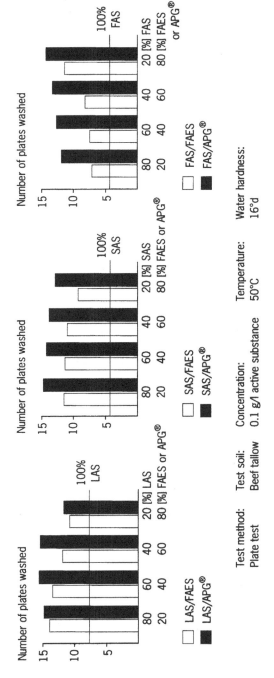

FIG. 17 Dishwashing performance of $C_{12/14}$-APG in combination with LAS, SAS, and FAS.

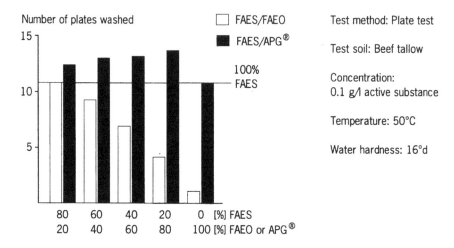

FIG. 18 Dishwashing performance of $C_{12/14}$-APG compared with FAEO in combination with FAES.

factants with cosurfactants. The combination of alkyl polyglycosides with betaine has been particularly beneficial in performance evaluation using mixed soil [52].

(b) Foaming Behavior. A feature of many conventional nonionic surfactants, such as FAEO, is their relatively low foaming capacity, alone or in combination with conventional anionic surfactants. In contrast, alkyl polyglycosides, which, when combined with anionic surfactants, show favorable foaming behavior, i.e. increase the foam volume or keep it at a high level. Figure 20 shows by way of example the influence of $C_{12/14}$-APG on the foaming capacity of FAS and FAES.

(c) Dermatological Behavior. Manual dishwashing detergents belong to the category of products which very frequently come into contact with the skin of the consumer, albeit in dilute form. Accordingly, the skin compatibility of this group of products is of particular interest. Alkyl polyglycosides are not only mild on the skin, but can significantly reduce the skin irritation of anionic surfactants. Initial indications of the skin compatibility of surfactants or surfactant combinations on a direct comparison basis can be obtained by a patch test [53]. The concentration of the test solutions is intentionally selected to produce a skin reaction, albeit slight. The degree of change in the skin is evaluated according to the following criteria: erythema, edema, squamation, and fissures, on a predetermined points scale.

The test results obtained with the primary surfactants LAS, SAS and FAS and

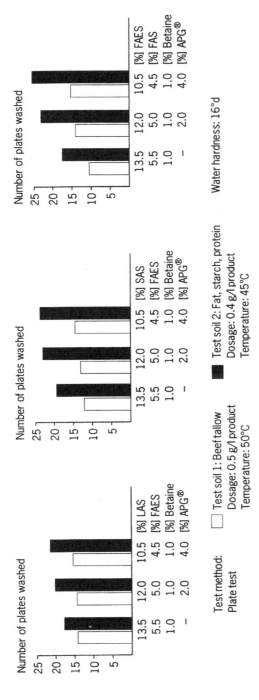

FIG. 19 Dishwashing performance of manual dishwashing detergents based on LAS, SAS, or FAS and FAES, betaine, and $C_{12/14}$-APG.

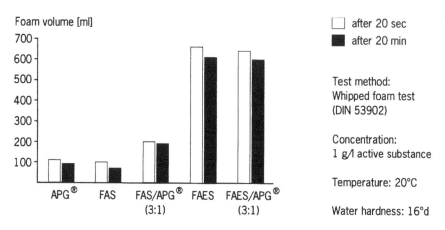

FIG. 20   Foaming behavior of $C_{12/14}$-APG-containing surfactant mixtures.

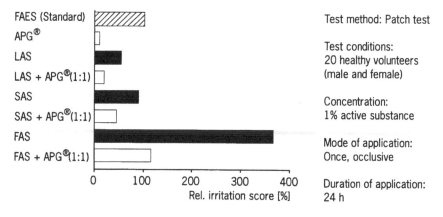

FIG. 21   Skin compatibility of $C_{12/14}$-APG and binary surfactant combinations.

combinations thereof with alkyl polyglycosides are set out in Figure 21. In every case, a distinct reduction in the relative total irritation scores is achieved when alkyl polyglycoside is combined with LAS, SAS, or FAS at same active substance content. Equally positive results are obtained in the in vitro test on the chorionallantois membrane on fertilized hens' eggs (HET-CAM) (Fig. 22) for determining mucous membrane compatibility [54]. Whereas hardly any improvement in mucous membrane compatibility is obtained by addition of betaines, a significant improvement is obtained by increasing additions of alkyl polyglycosides.

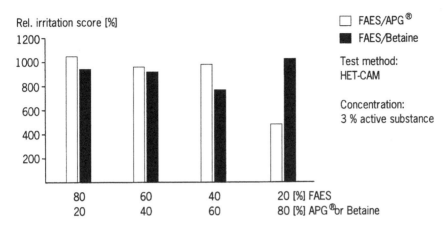

FIG. 22 Mucous membrane compatibility of $C_{12/14}$-APG compared with betaine in combination with FAES.

2. Alkyl Polyglycoside–Containing Manual Dishwashing Detergents in Europe

Within the group of hard-surface cleaners, MDDs are the most important in terms of tonnage. Thus, around 1.3 million tons of manual dishwashing detergents were produced in Europe in 1992 [55]. At the present, in Europe there are three segments: conventional manual dishwashing detergents, or high active manual dishwashing detergents (concentrates) and manual dishwashing detergents with excellent skin compatibility.

The most significant segment today is still that of conventional manual dishwashing detergents. However, they are by no means a homogeneous group from the formulation point of view. On the contrary, they may be divided into three groups (Table 3) which differ significantly in their active substance content and hence in their performance (for the same dosage). Products of relatively low concentration are found in Southern Europe while products of relatively high concentration are found, for example, in Great Britain. Since their introduction in Europe (Spain: 1984, Germany: 1992), very high active products that are used at half or one third the dose of conventional dishwashing detergents have experienced continual growth. Manual dishwashing detergents with excellent skin compatibility for consumers with sensitive skin were first introduced onto the European market by two manufacturers of branded goods at the end of 1992. Alkyl polyglycoside—containing MDDs which satisfy all consumer requirements are now available for all three market segments.

(a) Dishwashing Performance. The core property of a manual dishwashing detergent, namely its high cleaning performance, is still demanded by the con-

TABLE 3 Segmentation of Conventional Manual
Dishwashing Detergents

Group	Surfactants (wt.-%)	Countries
1	10–15	Portugal, Spain
2	15–27	Austria, France, Germany, Italy, Switzerland
3	35–40	United Kingdom

TABLE 4 General Formulations for Conventional and Concentrated
Manual Dishwashing Detergents of the European Market

Ingredients		wt.-%
Anionic surfactants	Lin. alkylbenzene sulfonate	10–35
	Sec. alkane sulfonate	
	Alkyl sulfate	
	Alkyl ether sulfate	
Nonionic surfactants	Alkyl polyethylene glycol ether	<15
	Alkyl polyglycoside	
	Fatty acid glucamide	
	Fatty acid alkanolamide	
	Alkyl dimethyl amine oxide	
Amphoteric surfactants	Alkylbetaine	<5
	Alkylamidobetaine	
Minor ingredients	Protein	<2
	Polymer	
Hydrotropes		<10
Fragrances		<1
Preservatives		<0.1
Colorants		<0.1
Salts		<2
Water		Balance

sumer. Table 4 illustrates the formulation scope available to the product developer. In Europe today, all three primary surfactant systems (LAS/FAES-, SAS/FAES-, FAS/FAES-based) are represented in the market both in conventional manual dishwashing detergents and in the concentrates.

The synergistic interactions of alkyl polyglycosides with the various primary surfactant systems enables the product developer to formulate even more effective products for the same active substance content or to reduce the active substance content without affecting the performance level (Fig. 23).

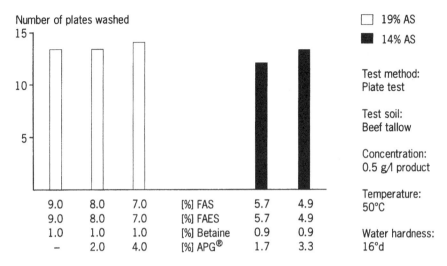

9.0	8.0	7.0	[%] FAS	5.7	4.9
9.0	8.0	7.0	[%] FAES	5.7	4.9
1.0	1.0	1.0	[%] Betaine	0.9	0.9
–	2.0	4.0	[%] APG®	1.7	3.3

FIG. 23 Dishwashing performance of conventional manual dishwashing detergents containing $C_{12/14}$-APG.

(b) Foaming Behavior. Although foam volume and foam structure both in the dishwashing liquor and under running water do not directly determine product performance, they are experienced by the consumer and thus lead to purchasing decisions. In alkyl polyglycoside, the product developer has a nonionic surfactant that improves the foam properties of the product. Both foam height and foam stability under mechanical influence can be investigated by the stress stability of foam test. In this test, the foam is produced by introducing air into the surfactant solution in defined quantities through a sieve of predetermined mesh width. The foams formed can be observed in regard to appearance and volume as a function of time. Additional information can be obtained by subjecting the foam thus produced to mechanical impact. Information on the sensitivity of the foams to fatty soil can be acquired by adding the soil at the beginning of the test and including it in the foaming process. The test results obtained with various commercial concentrates, even in the presence of soil, are set out in Figure 24. In this case, a mechanical impact was applied to the foam after 10 min and the foam height was determined in the following 5 min. The results show that the alkyl polyglycoside containing dishwashing detergent produces an elastic foam which is relatively unaffected by soil.

(c) Dermatological Properties. The product developer obtains initial indications of the skin compatibility of manual dishwashing detergents by the patch test. This enables him to test different formulations at the same time. In general, further dermatological studies are subsequently carried out under conditions of

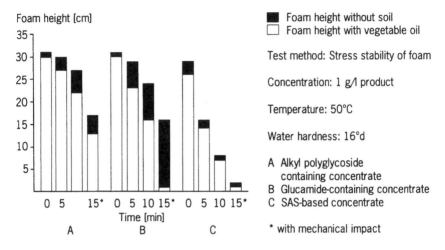

FIG. 24 Foaming behavior of concentrated manual dishwashing detergents.

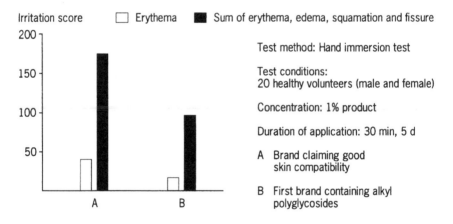

FIG. 25 Skin compatibility of conventional manual dishwashing detergents.

greater practical relevance. They include the hand immersion test [56] in which the consumers (at least 20 volunteers) every day immerse their hands in dishwashing liquid solutions under controlled conditions for several days. A dermatologist evaluates the degree of changes in the skin according to various criteria. When the first alkyl polyglycoside containing manual dishwashing detergent was nationally introduced in Europe in 1989, the hand immersion test was used to investigate the skin compatibility of the product in comparison with a marketed product claiming high skin compatibility. Figure 25 shows that the alkyl polyglycoside containing manual dishwashing detergent has significant advantages. In particular, it is distinguished by a distinct reduction in the formation of erythema and squamation. Nevertheless, there are significant numbers of consumers who still have problems using these conventional manual dishwashing detergents, so that they often wear gloves. For these consumers with sensitive and dry skin, products have been available since the end of 1992 which are demonstrably superior in skin compatibility to the products known up to that time. One product contains large quantities of nonionic surfactants (FAEO with a particular alkyl chain length and a special degree of ethoxylation) while the other is an anionic-surfactant-based formulation containing alkyl polyglycosides and other skin-protective components.

Figure 26 shows the results of a dermatological study on a group of volunteers with normal skin. The alkyl polyglycoside containing product has the best skin compatibility of all the products tested. Further studies, including the hand immersion test with a selected group of volunteers, proved that these products are also tolerated significantly better than conventional and concentrated manual dishwashing detergents, even by volunteers with dry and sensitive skin.

(d) Ecological Compatibility. As shown in Table 4, surfactants are by far the most important group of active substances in manual dishwashing detergents. Accordingly, they play a key role in the ecological compatibility of a dishwashing detergent. In many European countries, a minimum biological degradation rate is legally stipulated. However, the object of product development should be

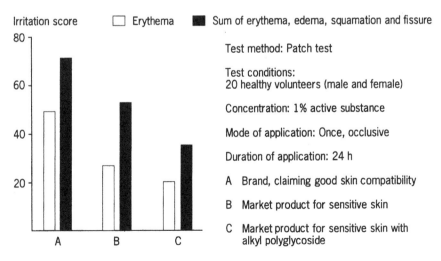

FIG. 26 Skin compatibility of conventional manual dishwashing detergents as compared to manual dishwashing detergents with excellent skin compatibility.

TABLE 5 General Formulations of Conventional Manual Dishwashing Detergents
with Similar Performance

Ingredients	MDD without $C_{12/14}$ APG (wt.-%)	MDD with $C_{12/14}$ APG (wt.-%)
Sec. alkane sulfonate	15–25	—
Alkyl sulfate	—	2–10
Alkyl ether sulfate	2–10	5–15
Alkyl polyglycoside	—	1–5
Alkylamidobetaine	—	1–5
Ethanol	<10	<10
Fragrances	<1	<1
Colorants	<0.1	<0.1
Water	Balance	Balance
Total amount of surfactants	25	20

to use raw materials with total biodegradability, i.e., proven ultimate biodegradation, preferably raw materials which are degraded not only aerobically but also
anaerobically. Alkyl polyglycosides are distinguished by optimal toxicological
and ecological data. The combination of alkyl polyglycosides with other surfactants characterized by very good biodegradability leads to products which fully
satisfy the requirements of consumers and authorities in regard to the ecological
compatibility of this product category.

Table 5 illustrates how high-performance commercial dishwashing detergents
can be formulated with alkyl polyglycosides for a significantly lower surfactant
content at the same time. The following estimation illustrates the environmental
impact of reducing the surfactant content of formulations. In 1992, around
300,000 tons of surfactants were used in manual dishwashing detergents in Europe. If high-quality products containing 20% lower active substance were to be
brought onto the market, the discharge of surfactants into domestic wastewater
would be reduced by 60,000 tons.

3. Alkyl Polyglycoside–Containing Manual Dishwashing Detergents in America

The products used in North America and in Latin America differ distinctly in
their composition and consistency [57].

(a) North America. In North America, there are at present four manual dishwashing detergents segments: conventional manual dishwashing detergents (private label and premium products), concentrates, manual dishwashing detergents
with excellent skin compatibility, and manual dishwashing detergents containing
an antibacterial agent.

The most important segment today is that of conventional manual dishwashing detergents (premium products) which are essentially based on two surfactant systems: LAS/FAES/fatty-acid alkanolamide and FAES/betaine/amine oxide. The inexpensive private label products (economy brands) have distinctly lower active substance contents and, accordingly, are of significantly lower performance for the same dosage. Nearly crystal-clear manual dishwashing detergents with excellent skin compatibility were introduced onto the market in 1992. They are based on FAEO and betaine and contain relatively small quantities of anionic surfactants, for example FAES. In 1995, various manufacturers began marketing manual dishwashing detergents which can also be used as a liquid hand soap and which contain an antibacterial agent. In North America, too, the advantages of alkyl polyglycosides (performance, foaming behavior, and skin compatibility) have already resulted in the reformulation of more than 10 brands.

Conventional manual dishwashing detergents. The majority of conventional dishwashing detergents are based on LAS due to the relatively high cost of FAES as compared with LAS. As in Europe, alkyl polyglycosides are used as co-surfactants in these products. The optimum cost/effectiveness ratio of LAS/alkyl polyglycoside surfactant combinations under American conditions is between 3:1 and 4:1. Formulations and selected performance figures both for LAS- and FAES-based products are shown in Figure 27. The performance of both the LAS- and the FAES-based premium product is comparable with that of the market leader in the premium segment. Various manufacturers have meanwhile introduced such alkyl polyglycoside containing products onto the market.

Concentrated manual dishwashing detergents. Concentrated manual dishwashing detergents known as ultraliquids were brought onto the market in 1995 by various manufacturers. These products are marketed in 14.7-oz bottles in contrast to the conventional 22-oz bottles. The recommended dosage is one-third lower than that of conventional products for the same performance. Ultra liquids contain around 45% to 50% surfactants as compared to about 30% in conventional dishwashing detergents. Accordingly, the use of alkyl polyglycosides is an advantage in ultraliquids because, by virtue of the interactions with anionic surfactants described earlier, highly effective products with relatively low active substance contents can be formulated with alkyl polyglycosides. In addition, alkyl polyglycoside containing ultra liquids require less hydrotrope than non–alkyl polyglucoside ultraliquids. As in the case of conventional dishwashing detergents, the partial or complete replacement of FAES by alkyl polyglycosides leads to increases in performance. Table 6 contains examples of alkyl polyglycoside containing ultra liquids which are comparable in cost and performance with leading ultra liquids of the North American market.

I&I solid blocks. In addition to liquid manual dishwashing detergents, so-called solid blocks are also available on the American I&I market. Doses of these products are dispensed by means of a jet of water which dissolves out part

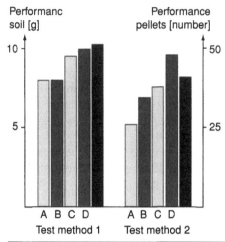

Performanc soil [g]

Performance pellets [number]

■ Market leader in the premium category

Test method 1: Tergotometer
Test soil: Fat, protein on cotton swatches
Concentration: 1.5 g/1 product
Temperature: 110°F (43°C)
Water hardness: 150 ppm (Ca:Mg=3:2)

Test method 2: Modified Shell
Test soil: Fat, protein, starch
Concentration: 40 g/1 product
Temperature: 110°F (43°C)
Water hardness: 150 ppm (Ca:Mg=3.2)

Ingredients	Economical manual dish-washing detergents based on		Premium dish-washing detergents based on	
	LAS/APG®	FAES/APG®	LAS/APG®	FAES/APG®
	A	B	C	D
Lin. alkylbenzene sulfonate	12.6	–	19.2	–
Alkyl ether sulfate	–	9.8	–	14.2
Alkyl polyglycoside	3.2	3.2	4.8	4.6
Alkylamidobetaine	1.0	5.0	2.0	7.2
Fatty acid dialkanolamide	1.0	–	2.0	–
Ethanol	1.8	–	3.2	–
Xylene sulfonate	1.6	2.5	3.2	6.5
Water	Balance	Balance	Balance	Balance

FIG. 27 Conventional $C_{12/14}$-APG-containing manual dishwashing detergents in North America.

of the solid block and carries it into the dishwashing liquor. Apart from performance aspects like fat dissolving power and foaming behavior, in the case of solid blocks, processing, hardness, and dissolving behavior are also important. Effective alkyl polyglycoside–containing formulations (Table 7) with favorable dissolving behavior can be provided for this segment also. The hardness of the blocks can be controlled through the ratio of fatty-acid monoethanolamide to fatty-acid diethanolamide.

(b) *Latin America.* In Latin America, there are three segments to the manual dishwashing detergent market: highly viscous, liquid manual dishwashing detergents of relatively low concentration; low active low viscous manual dishwash-

TABLE 6 $C_{12/14}$ APG-Containing Ultraconcentrates in North America

Ingredients	LAS/APG (wt.-%)	FAES/APG (wt.-%)	FAES/APG (wt.-%)
Lin. alkylbenzene sulfonate	24.0	—	—
Alkyl ether sulfate	—	28.0	30.0
Alkyl polyglycoside	6.0	7.2	10.0
Fatty alcohol ethoxylate	8.0	6.0	—
Alkylamidopropylbetaine	3.2	—	—
Fatty acid dialkanolamide	3.0	3.0	6.0
Fatty acid amidopropyl amine oxide	—	3.0	—
Sodium chloride	—	5.0	—
Ethanol	—	6.0	8.0
Xylene sulfonate	6.0	—	—
Water	Balance	Balance	Balance

TABLE 7 $C_{12/14}$ APG-Containing Solid Block Dishwashing Formulations in North America

Ingredients	Solid block 1 (wt.-%)	Solid block 2 (wt.-%)	Solid block 3 (wt.-%)
Lin. alkylbenzene sulfonate	13.2	13.2	13.2
Alkyl ether sulfate	12.0	12.0	12.0
Fatty acid	6.0	6.0	6.0
Alkyl polyglycoside	4.0	4.0	4.0
Fatty acid monoalkanolamide	10.0	—	5.0
Fatty acid dialkanolamide	—	10.0	5.0
Nonyl phenol ethoxylate	4.5	4.5	4.5
Alkylamidobetaine	6.0	6.0	6.0
Ethanol	3.0	3.0	3.0
Urea	20.0	20.0	20.0
Sodium hydroxide	0.75	0.75	0.75
Water (from raw materials)	Balance	Balance	Balance

ing detergents; and paste-form manual dishwashing detergents which, besides surfactants, can also contain builders, abrasives, and fillers. In Latin America, too, the advantages of alkyl polyglycosides (performance, foaming behavior, skin compatibility, and increase in viscosity in products of relatively low concentration) have also resulted in the introduction of alkyl polyglycoside containing products.

Liquid manual dishwashing detergents.   Consumers in some Latin American

TABLE 8 Manual Dishwashing Detergents in Latin America

Ingredients	Premium product (wt.-%)	Economy product (wt.-%)
Lin. alkylbenzene sulfonate	12.6	4.8
Alkyl ether sulfate	—	3.75
Alkyl polyglycoside	4.0	1.25
Fatty acid dialkanolamide	1.0	3.0
Alkylamidobetaine	1.0	—
Ethanol	—	1.0
Sodium chloride	0.25	1.0

countries prefer high-viscosity products (around 1000 mPa·s) with high foaming behavior. Typical market products are based on LAS or on combinations of LAS/FAES and contain 10% to 15% of active substances. Table 8 shows typical formulations which satisfy Latin American requirements both in regard to viscosity and performance.

Manual dishwashing pastes. Many Latin American markets are dominated by dishwashing pastes which are now available in two forms: conventional pastes containing solids, and the new clear pastes, which contain no solids. The conventional pastes consist of LAS, abrasives (calcium carbonate), builders (sodium tripolyphosphate), hydrotropes, and fillers. The use of alkyl polyglycosides in these products significantly improves both dishwashing performance and foaming behavior. The performance of an alkyl polyglycoside containing paste is compared with that of a market product in Figure 28. In the alkyl polyglycoside–containing paste, 2% of the active LAS in the market product was replaced by 1% of the active FAES: $C_{12/14}$-APG (3:1). The results prove that even the use of small quantities of APG can lead to a distinct improvement in performance. The clear pastes are also based on LAS. They contain large quantities of hydrotropes. Examples of alkyl polyglycoside containing pastes which may be both transparent and opalescent in appearance are given in Table 9. The hardness of the pastes can be controlled either through the pH value (low pH value = soft paste) or through the concentration of fatty acid. In addition, an LAS with a high content of 2-phenyl isomers is required for the production of opalescent pastes.

4. Alkyl Polyglycoside–Containing Manual Dishwashing Detergents in Asia

In the Asian countries, consumers are concerned mainly with the dermatological properties of manual dishwashing detergents; in addition, there should be no risk to health if foods (for example, fruit and vegetables) come into contact with these products. The demand for products highly compatible with the skin is un-

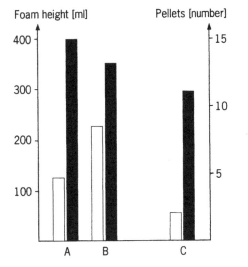

Foam height [ml]   Pellets [number]   ☐ Market paste
■ $C_{12/14}$ APG® containing paste

Test method A: Sponge foam
Concentration: 1 g/l product
Temperature: room temperature
Water hardness: 50 ppm (Ca:Mg=2:1)

Test method B: Inverted cylinder
Concentration: 0.5 g/l product
Test soil: Fat, protein, starch
Temperature: room temperature
Water hardness: 150 ppm (Ca:Mg=3:2)

Test method C: Modified Shell
Test soil: Fat, protein, starch
Concentration: 40 g/l product
Temperature: 110°F (43°C)
Water hardness: 150 ppm (Ca:Mg=3:2)

FIG. 28   Performance of dishwashing pastes.

TABLE 9   Opalescent and Transparent Pastes Containing $C_{12/14}$ APG

Ingredients	Opalescent paste (wt.-%)	Transparent paste (wt.-%)
Lin. alkylbenzene sulfonic acid	6.7	10.0
Alkyl ether sulfate	—	10.5
Fatty acid	10.0	4.0
Alkyl polyglycoside	5.0	4.0
Nonyl phenol ethoxylate	17.5	—
Fatty acid dialkanolamide	15.0	4.0
Ethanol	—	1.0
Isopropanol	5.0	3.0
Polyethylene glycol 400	—	10.0
Urea	20.0	20.0
Sodium hydroxide	4.3	3.7
Water	Balance	Balance

derstandable from the different dishwashing customs in most Asian countries as compared with Europe and North America. Whereas in Europe dishes are mainly washed in dilute detergent solutions (~ 0.8 to 4.0 g of product/L wash liquor), the consumer in Asian countries usual washes up with a sponge impregnated with the concentrated detergent. Accordingly, this method of dishwashing leads to far more direct skin/detergent contact than in Europe. In order to meet the stringent skin compatibility requirements, manual dishwashing detergents with high contents of nonionic surfactants are produced in such countries as Japan, Korea, Hong Kong, and Taiwan and recently in Malaysia and Thailand. FAEO, alkyl polyglycosides, fatty-acid alkanolamides, amine oxides, and also betaines are preferably used as the nonionic surfactants. The first alkyl polyglycoside–containing conventional dishwashing detergent was introduced onto the Japanese market in 1989. In the meantime, alkyl polyglycoside–containing concentrates have also been brought onto the market. In Asian countries, in contrast to Europe and North America, alkyl polyglycoside is also used as a primary surfactant and not only as a cosurfactant.

In Indonesia, the Philippines, Malaysia, and Thailand, in addition to the conventional low-concentration dishwashing detergents based on linear or branched alkyl benzene sulfonate, there are creamy products for the general cleaning of hard surfaces. Here also skin compatibility can be further optimized by the use of alkyl polyglycosides.

B.   Alkyl Polyglycosides in Cleaners

The relatively long-chain alkyl polyglycosides with an alkyl chain length of $C_{12/14}$ and a DP of around 1.4 have proved to be of particular advantage for manual dishwashing detergents. However, the relatively short-chain alkyl polyglycosides with an alkyl chain length of $C_{8/10}$ and a DP of about 1.5 ($C_{8/10}$-APG) are particularly useful in the formulation of all-purpose and specialty cleaners. Formulations for cleaners containing surfactants and surfactant combinations based on petrochemical and vegetable feedstocks are sufficiently well known. Extensive knowledge has been built up on this subject [58]. Now that light-colored short-chain alkyl polyglycosides are also available on the market, many new applications are being found for alkyl polyglycosides by virtue of their broad performance spectrum:

Good cleaning efficiency
Low environmental stress cracking potential (ESC) for plastics
Transparent residues
Good solubility
Good solubilization
Stability against acids and alkalis
Improvement of low temperature properties of surfactant combinations
Low skin irritation
Excellent ecological and toxicological properties

Today, alkyl polyglycoside–containing products are found both in all-purpose cleaners and in special cleaners, such as bathroom cleaners, toilet cleaners, window cleaners, kitchen cleaners, and floor care products [59].

1. All-Purpose Cleaners

The broad range of soil types found in the home require modern all-purpose cleaners (APC). They have to perform effectively against both emulsifiable oil- and fat-containing soils and against dispersible solid soil particles. Today, there are three segments for the European all-purpose cleaner market: conventional all-purpose cleaners, concentrated all-purpose cleaners, and all-purpose cleaners with excellent skin compatibility. Table 10 illustrates the formulation scope for the product developer. Essential ingredients are surfactants and builders. Alkyl polyglycoside containing products are now available on the market for all three segments. Alkyl polyglycosides themselves have an excellent cleaning performance which can be determined, for example, in accordance with the IPP quality standard [60]. The cleaning performance can be further increased by small additions of anionic surfactants and/or polymeric boosters.

Thus, it is possible to formulate products comparable in cleaning performance to the market leaders at significantly lower surfactant contents. All-purpose cleaners with particularly good skin compatibility should be slightly acidic rather than alkaline. With alkyl polyglycoside, the product developer has a sur-

TABLE 10   General Formulations for Conventional and Concentrated All-purpose Cleaners

Ingredients		Conventional APC (wt.-%)	Double conc. APC (wt.-%)
Surfactants	Lin. alkylbenzene sulfonate	5–10	10–20
	Sec. alkane sulfonate		
	Alkyl sulfate		
	Alkyl ether sulfate		
	Soap		
	Alkyl polyalkylene glycol ether		
	Alkyl polyglycoside		
Builders	Citrate	1–2	1–3
	Gluconate		
	Bicarbonate/carbonate		
Solvents/hydrotropes	Cumene sulfonate	0–5	1–6
	Alcohol		
	Glycol ether		
Additives	Fragrances	<1	<2
	Colorants		
	Preservatives		
Water		Balance	Balance

factant of which the high cleaning performance level is hardly affected by changes in the pH value.

Consumers today prefer all-purpose cleaners with moderate or low foaming behavior. The foaming capacity of alkyl polyglycoside–containing cleaners can readily be reduced by using small quantities of soaps or increased by adding small quantities of anionic surfactants. A suitable foaming capacity can thus be adjusted for each country. Alkyl polyglycoside has proved to be the problem solver in the formulation of concentrated all-purpose cleaners with excellent ecological compatibility. With alkyl polyglycosides, it is possible to formulate concentrates which have a correspondingly higher content of builders and perfume oils and require lower quantities of hydrotropes.

2. Bathroom Cleaners

Bathroom cleaners today are used in the form of liquid and foam pumps or aerosol packs. Table 11 shows the formulation scope for liquid bathroom cleaners. Bathroom cleaners are generally adjusted to an acidic pH although it is important to ensure that damage to sensitive enamels is avoided. A pH range of 3 to 5 is recommended for bathroom cleaners. The most common soils in bathrooms are greasy soils, lime soap residues and lime residues based on the hardness of the tap water. The performance of a cleaner consisting of 4% surfactant, 4% citric acid/citrate, and 2.5% ethanol against these soils is shown in Figure 29. The

TABLE 11 General Formulations for Bathroom Cleaners

Ingredients		wt.-%
Surfactants	Lin. alkylbenzene sulfonate	2.0–8.0
	Sec. alkane sulfonate	
	Alkyl sulfate	
	Alkyl ether sulfate	
	Alkyl polyethylene glycol ether	
	Alkyl polyglycoside	
Builder	Citrate, etc.	0.5–2.0
Acids	Citric acid	3.0–6.0
	Acetic acid	
	Lactic acid	
	Dicarboxylic acid	
Solvents	Alcohols	2.0–9.0
	Glycols	
Additives	Fragrances	<1
	Colorants	
	Preservatives	
Water		Balance

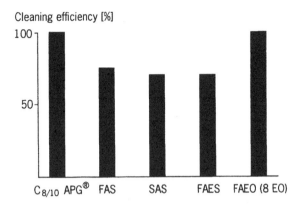

Cleaning efficiency [%]

Test method:
IPP quality norm

Test soil:
Oil, carbon black,
inorganic pigments, paraffins

Concentration:
Concentrated (spray product)

Temperature:
20°C

Water hardness:
16°d

FIG. 29 Cleaning efficiency according to IPP.

TABLE 12 Plastics in Bathrooms

Plastic sensitve to environmental stress cracking	Plastic less sensitive to environmental stress cracking
Acrylonitrile, butadiene, styrene	Cellulose propionate
ABS/polycarbonate	Polyamide
Polycarbonate	Polymethylene, polyformaldehyde
Polymethacrylate	Polypropylene

surfactant components investigated were $C_{8/10}$-APG, FAEO (8 EO), FAS, SAS, and FAES. Another important performance feature of bathroom cleaners besides their actual cleaning performance is the avoidance of environmental stress cracking in components made of plastic, including handles, water overflows in bath tubs, fitting, or shower holders. Important bathroom plastics are listed according to their sensitivity to environmental stress cracking (ESC) in Table 12. By interacting with the various ingredients present in bathroom cleaners, plastics can develop environmental stress cracks which, in the past, have resulted in considerable claims against the manufacturers. In the laboratory, environmental stress cracking effects can be simulated both by the pin impression method and by the ben strip method [61]. In the pin impression method, the plastic is prestressed by indentation with a steel pin and is then immersed in the test solution for 15 min. The plastic is then dried and evaluated after 24 hours.

In the ben strip method, plastic strips are prestressed by weights. The test solution is then allowed to act on the plastic strips for 15 min. The strips are then dried and likewise evaluated after 24 hours. The results obtained with low-concentration

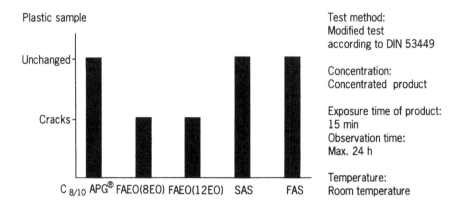

FIG. 30   Environmental stress corrosion of bathroom cleaners.

bathroom cleaners, for example of the aerosol type, are set out in Figure 30. $C_{8/10}$-APG as a nonionic surfactant is as favorable in regard to preserving behavior as conventional anionic surfactants.

3.   Window Cleaners

Modern window cleaners consist mainly of a surfactant component and a solvent component (Table 13). Besides a good wetting effect on the soil and the glass surfaces, the surfactants must also have a good fat-emulsification power. In addition, the surfactant residue on the surface must be transparent and easily removable by polishing.

Normal use provides for the direct removal of soil in a single operation, i.e. spraying on, spreading the liquid with a clean cloth, then removing the liquid together with the detached soil with a view to obtaining a clean and streak-free window. If not properly applied, the product can dry and individual non-volatile components can remain behind in the form of troublesome coatings and greasy films.

With reference to the example of formulations containing 0.1% to 0.4% surfactant, 1% to 3% glycol ether, and > 1% perfume, it is shown that purely anionic surfactant-containing residues are more difficult to polish out than combinations of anionic surfactants with nonionic surfactants (Fig. 31). The residue of an FAS/$C_{8/10}$-APG-containing formulation can be polished out particularly effectively. The optimum ratio of FAS to alkyl polyglycoside depends on the perfume oil used and upon the associated total quantity of surfactant. Besides good polishability, the FAS/$C_{8/10}$-APG combination also shows advantages in favorable foaming behavior and very good solubilization of perfume oil. Good foaming behavior is generally difficult to obtain where butoxyethanol is used as

TABLE 13 General Formulations for
Window Cleaners

Ingredients	wt.-%
Surfactants	0.05–0.4
Olefin sulfonate	
Sec. alkane sulfonate	
Alkyl sulfate	
Alkyl ether sulfate	
Alkyl polyglycoside	
Solvents	3.0–20.0
Ethylene glycol ether	
Propylene glycol ether	
Ethanol	
Isopropanol	
Additives	<1
Fragrances	
Colorants	
Water	Balance

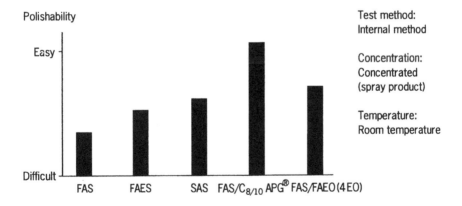

FIG. 31 Polishability of residues.

the solvent. Figure 32 shows that the use of $C_{8/10}$-APG is also advantageous in this case.

C. Alkyl Polyglycosides in Laundry Detergents

For many consumers, laundry detergents are a product they use daily to bring soiled clothing back into a state fit for use. The necessary formulations are

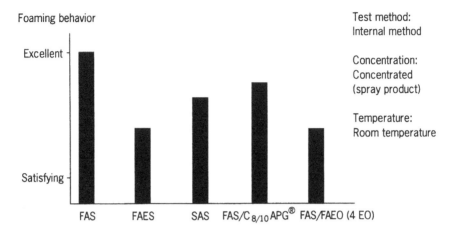

FIG. 32   Foaming behavior of window cleaners.

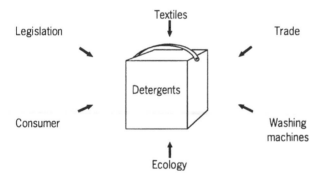

FIG. 33   Influences on the development of detergents.

marketed in various forms, for example as powder, paste, or liquid detergents. The choice of the particular formulation is determined by the type of soil, by consumer requirements in regard to ease of use, and last but not least, by the textile and its washing instructions. In addition, ecology has been an important factor in the development of laundry detergents, influencing the way in which they are developed.[62] Apart from legislative measures, voluntary agreements by the industry, and commercial preferences, the consumer ultimately makes the decision for or against the purchase of a certain product (Fig. 33).

The alkyl polyglycosides used in detergent formulations are those with an alkyl chain length of $C_{12/14}$ and a DP of around 1.4 ($C_{12/14}$-APG). As a nonionic surfactant, they are particularly effective against fatty soils. Optimized surfactant systems generally based on mixtures of anionic and nonionic surfactants are used in modern detergent formulations. Alkyl polyglycosides occur in these surfactant mixtures preferably as so-called cosurfactants which have the property of complementing or improving the quantitatively predominant main surfactants in regard to washing performance. Besides performance, the aesthetics of a detergent plays an important part. Wool detergents, for example, are intended to produce a rich, stable foam. The consumer associates performance and care with foam. For machine washing, higher foam height on the one hand improves textile care because the mechanical action on the wash load is reduced. On the other hand, this may cause a distinct reduction of the detergency performance. Alkyl polyglycosides in conjunction with anionic surfactants may alter the foaming behavior of the formulations. A general observation on foaming behavior is not possible, but is to a large extent dependent on the surfactants used and their ratios to one another.

1. Liquid Detergents

Alkyl polyglycosides were first used in liquid laundry detergents in 1989. The composition of a liquid heavy-duty detergent (Table 14) is based on a combination of nonionic surfactants, anionic surfactants, soaps, and hydrotropes. The hydrotropes—which do not contribute to cleaning—can be partly replaced by alkyl polyglycosides. It has surprisingly been found that alkyl polyglycosides positively influence the low-temperature and storage stability of such formulations. In addition, triethanolamine (TEA) soaps have been successfully replaced by the similarly acting sodium/potassium soaps.[63] The diethanolamine contamination

TABLE 14 General Formulations of Liquid Detergents

Ingredients	Heavy-duty liquid detergent concentrate (wt.-%)	Heavy-duty liquid detergent (wt.-%)	Light-duty liquid detergent (wt.-%)
Anionic surfactants	4–9	4–9	4–9
Nonionic surfactants	20–40	10–20	10–15
Soap	10–20	10–20	4–9
Alkyl polyglycoside	1.5–3	0.5–1.5	4–9
Ethanol	4–9	3–8	3–8
Glycerine	4–9	3–8	—
Optical brightener	+	+	—
Enzymes (protease, amylase)	+	+	—
Fragrances, pearlescence	+	+	+
Water	Balance	Balance	Balance

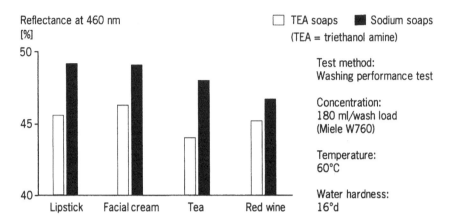

FIG. 34 Washing performance of heavy-duty liquid detergents.

of triethanolamine, which may cause the formation of nitrosamines, was thus avoided. In addition, sodium soaps are less expensive so that, as a net result, the use of alkyl polyglycosides may give the formulation a price advantage. Washing tests show that these new formulations outperform their predecessors (Fig. 34). Whereas in the case of liquid heavy-duty detergents containing optical brighteners and enzymes, the emphasis is on performance, the care aspect is more in the foreground in the case of specialty detergents. The textile influence is dominant in this regard. Liquid specialty detergents are formulated having pH values of < 8.5. In order to avoid shifts in color, optical brighteners are not used in the formulations.[64]

Also consumers expect such products to have an appearance comparable with cosmetic formulations, such as hair shampoos. By using alkyl polyglycosides, it is possible to fulfill these wishes [65]. Thus, the viscosity of the liquid is increased by the use of $C_{12/14}$-APG (Fig. 35) [66]. These formulations are simple and safe to handle and for handwashing are preferred to powders.

The storage stability of enzymes in liquid formulations is reduced when compared with powders. On account of the high surfactant content of certain formulations, the enzymes are partly deactivated and slowly loose their initial activity upon storage. In order to improve the storage stability of enzymes, such as proteases, lipases, amylases and/or cellulases, in liquid detergents, stabilizers (borates, phosphates, special esters) are added and the surfactant systems are adapted. It has been found that the storage stability of enzymes in liquid detergents can be distinctly improved by the use of $C_{12/14}$-APG (Fig. 36). Furthermore, alkyl polyglycosides have the advantage over the otherwise typical stabilizers of contributing to the washing performance.

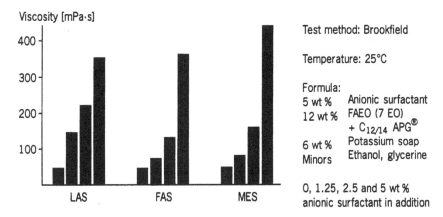

FIG. 35 Increase of viscosity of liquid detergents. LAS = linear alkylbenzene sulfonate; FAEO = fatty alcohol ethoxylate; FAS = fatty alcohol sulfate; MES = methyl ester sulfonate.

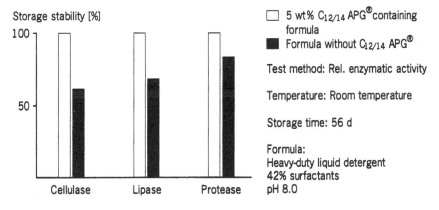

FIG. 36 Stability of enzymes in liquid detergents.

2. Powder Detergents

Quantitatively the largest group, the heavy-duty powder detergents are based on formulations which remove virtually all the soil types normally encountered (Fig. 37). Particular emphasis is placed on washing performance. For this reason, a distinctly higher alkalinity is adjusted so that the pH value of such detergents is in the range from pH 9.5 to 10.5. Soil removal is thus greatly improved. In addition, heavy-duty detergents are provided with a bleaching system. Bleachable stains, such as tea, coffee, red wine, etc., are thus effortlessly removed. Fat- and

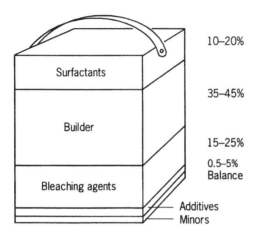

10–20%

35–45%

15–25%

0.5–5%
Balance

FIG. 37   Composition of a heavy-duty detergent.

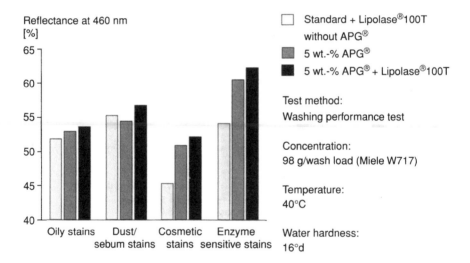

FIG. 38   Washing performance of heavy-duty detergents with $C_{12/14}$-APG.

oil-containing soils, such as sebum, olive oil, lipstick, and facial cream, are difficult to remove, particularly at low temperatures. By using alkyl polyglycosides
in powder-form detergents, these stains in particular can be removed considerably more effectively (Fig. 38). By additionally using lipases, washing performance can be further increased.

Henkel was the first German detergent manufacturer to use alkyl polyglycosides in a powder detergent. This detergent without bleach and optical brightener is used for fine and delicate textiles like silk and viscose. These high-priced textiles need a detergent which is more sensitive to fibres without any damaging during the wash cycle. It is possible to generate a rich and creamy foam by using alkyl polyglycosides. This microfoam reduces the mechanic action during the wash cycle with the big advantage of increased textile care. The product also has very good washing performance in cold water (below 40°C).

V. ALKYL POLYGLYCOSIDES IN PERSONAL-CARE PRODUCTS

Over the past decade, progress in the development of raw materials for personal-care products has mainly occurred in three areas: mildness and care for the skin; high-quality standards by minimization of byproducts and trace impurities; and ecological compatibility. At the same time, the variety of parameters to be tested for a new raw material has increased and their detection by subjective and objective methods has become more precise. Official regulations and the needs of the consumer have increasingly stimulated innovative developments which follow the principle of sustainability of processes and products. The use of renewable raw materials for the production of alkyl polyglycosides from vegetable oils and carbohydrates is one aspect of this principle. A life-cycle analysis has been published for the production of alkyl polyglycosides under European conditions providing an overview of the energy and resource demands and the environmental emissions involved in the production of 1000-kg $C_{12/14}$ alkyl polglycosides [67]. The development of a commercial technology requires a high level of control of the raw materials and the reaction and working-up conditions to satisfy modern quality requirements for cosmetic raw materials at reasonable cost.

A. Cosmetic Cleansing Formulations

In the cosmetic field alkyl polyglycosides represent a new class of surfactants which combine properties of conventional nonionics and anionics. By far the largest proportion of commercial products is represented by C_{8-14} alkyl polyglycoside for cosmetic cleansing formulations. Their composition (Table 15) is designed for optimum basic performance, such as foaming, viscosity, and handling properties.

1. Dermatological Properties

For body-cleansing formulations, a new modern surfactant must have excellent compatibility with the skin and mucous membranes. Dermatological and toxicological tests are essential for the risk assessment of a new surfactant and are designed above all to determine possible irritation of living cells in the basal layer

TABLE 15 Technical Data of Alkyl Polyglycosides[a]

Plantacare 818	Plantacare 1200	Plantacare 2000 Plantaren 1200	Plantaren 2000
INCI name[b]	Coco glucoside	Lauryl glucoside	Decyl Glucoside
Active substance (AS)	51–55%	50–53%	51–55%
pH value (20%)	11.5–12.5	11.5–12.5	11.5–12.5
Appearance (20%)	Liquid	Pasty	Liquid
Recommended storage temperature	>15°C	38–45°C	>0°C
Preservation	None	None	None

[a]See also Chapter 2, Table 1.
[b]Capital letters are used for INC names in the entire chapter.

of the epidermis (primary irritation). In the past, this was the basis for such claims as "mildness" of a surfactant. In the meantime, the meaning of mildness has changed considerably. Today, mildness is understood to be the all-round compatibility of a surfactant with the physiology and function of human skin—more precisely, the epidermis.

The physiological effects of surfactants on the skin are investigated by various dermatological and biophysical methods starting with the surface of the skin and progressing via the horny layer and its barrier function to the deeper layer of the basal cells. At the same time, subjective sensations, such as the feeling on the skin, are recorded by verbalization of tactile sense and experience.

Alkyl polyglycosides with C_8 to C_{16} alkyl chains belong to the group of very mild surfactants for body-cleansing formulations. In a detailed study, the compatibility of alkyl polyglycosides was described as a function of the pure alkyl chain and the degree of polymerization [68]. In the modified Duhring Chamber Test, C_{12} alkyl polyglycoside shows a relative maximum within the range of mild irritation effects, whereas C_8, C_{10}, C_{14}, and C_{16} alkyl polyglycosides produce lower irritation scores. This corresponds to observations with other classes of surfactants. In addition, irritation decreases slightly with increasing degree of polymerization (from DP = 1.2 to DP = 1.65).

On the other hand, mucous membrane compatibility, as determined by the in vitro test on the chorionallantois membrane of fertilized hens' eggs (HET-CAM) as an alternative to Draize's mucous membrane compatibility test, shows a monotonic decrease from C_8 to C_{14} alkyl polyglycoside within the range of mild irritation. The DP causes only a slight differentiation of compatibility in this case.

The commercial APG products (Plantacare/Plantaren 1200, Plantacare/Plantaren 2000, and Plantacare 818) with mixed alkyl chain lengths have the best overall compatibility with relatively high proportions of long-chain ($C_{12/14}$)

alkyl polyglycosides. They join the very mild group of highly ethoxylated alkyl ether sulfates, amphoglycinates, or amphodiacetates and the extremely mild protein fatty-acid condensates based on collagen or wheat protein hydrolyzates. Comparative tests have been carried out [69,70].

Another in vitro test, the Red Blood Cell Test, investigates the hemolysis of erythrocytes under the influence of surfactants. For alkyl polyglycosides effects are very low as for other very mild surfactants [71,72]. A $C_{12/14}$ alkyl polyglycoside was subjected to the RBC test. A steep concentration-response relationship was observed: a concentration of 80 µg/g did not induce any hemolysis whereas a concentration of only 120 µg/g produced total hemolysis. The H_{50} value was found to be 104 µg/g [73].

Similarly, alkyl polyglycosides produce a very mild skin reaction in the modified Duhring Chamber test [70,74,75]. In mixed formulations with standard alkyl ether sulfate (sodium laureth sulfate, SLES), the score for erythema decreases with increasing levels of alkyl polyglycoside without a synergistic effect being observed (Fig. 39).

A complete study of skin compatibility investigates surface conditions, the influence on the transepidermal water loss through the horny layer, dermatological parameters, and subjective sensorial effects after surfactant application. This compilation test system also enables combination effects to be studied so that most compatible formulations can gradually be built up from basic surfactants, additives, and special ingredients [75]. A suitable test procedure is the arm flex wash test [76], which simulates the daily use of a product by open repeated application under accelerated conditions [70,75,77].

A more detailed evaluation of a skin surface profile after application of alkyl polyglycoside is provided by profilometry. The various statistical parameters which differentiate roughness of the skin describe a skin smoothing effect for alkyl polyglycosides [71,75,78]. The standardized mean swelling values (Q ac-

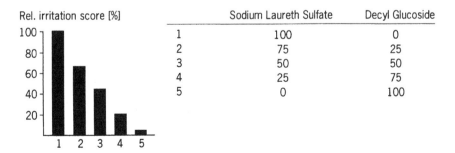

FIG. 39 Modified Duhring Chamber Test with relative irritation scores for erythema formation.

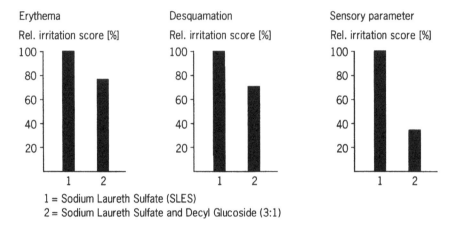

1 = Sodium Laureth Sulfate (SLES)
2 = Sodium Laureth Sulfate and Decyl Glucoside (3:1)

FIG. 40   Skin effects in the arm flex wash test: dermatological assessment.

cording to Zeidler) [79] of the horny layer, as determined on isolated pig epidermis, lie in the swelling-inhibiting range, i.e., less swelling than pure water. For $C^{12/14}$ alkyl polyglycoside, Q values of –6% to –9% (±7% with 95% confidence) were measured at pH 5.6.

The lower swelling of the epidermis by an alkyl polyglycoside solution as compared with water is a sign of the functional compatibility of alkyl polyglycosides and contributes by way of compensation to limiting irritation mechanisms of other components in the formulation. In addition, on the basis of corneometer measurements [74], the loss of moisture content in the horny layer is lower by 30% to 40% under the effect of alkyl polyglycoside as compared with standard ether sulfate. This corresponds to the effect of the very mild zwitterionic amphodiacetates.

The influence of surfactants on the barrier function of the epidermis either by deterioration of the functional structure or by elution of components (horny layer lipids, NMF) is characterized by evaporimeter measurements as a change of the natural transepidermal water loss (TEWL). Investigations in connection with the arm flex wash test show that the changes in relation to the normal state of the skin barrier produced by standard surfactants are reduced in the presence of alkyl polyglycosides. This effect can be increased in the systematic buildup of formulations by incorporating further additives, such as protein derivatives [70,75].

The dermatological findings in the arm flex wash test show the same ranking as in the modified Duhring Chamber Test where mixed systems of standard alkyl ether sulfate and alkyl polyglycosides or amphoteric co-surfactants are investigated. However, the arm flex wash test allows better differentiation of the effects. Formation of erythema and squamation can be reduced by 20% to 30% if ~25% of SLES is replaced by alkyl polyglycoside (Fig. 40).

Of greater importance is the recording of subjective feelings of the volunteers (itching, stinging, etc.), which indicates a reduction of about 60%. In the systematic buildup of a formulation, an optimum can be achieved by the addition of protein derivatives or amphoterics [75].

B. Performance Properties

1. Concentrates

Addition of alkyl polyglycosides modifies the rheology of concentrated surfactant mixtures so that pumpable, preservative-free, and readily dilutable concentrates containing up to 60% active substance can be prepared. These concentrated blends of several components may be generally used as cosmetic raw materials or specifically as core concentrates in the production of cosmetic formulations (for shampoos, shampoo concentrates, foam baths, shower gels, etc.).

Accordingly, the compositions with alkyl polyglycosides are based on highly active anionics, such as alkyl ether sulfate (sodium or ammonium salts), betaines, and/or nonionic surfactants and as such are more gentle on the eyes and the skin than conventional systems. At the same time, they show excellent foam behavior, thickening behavior, and processing properties. Super concentrates are preferred for economic reasons insofar as they are easier to handle and to dilute without containing hydrotropes. The mixing ratio of the surfactant base is adapted to the performance requirements of formulations.

A simple characteristic example is the compound of standard alkyl ether sulfate (for example, Texapon N 70) and $C_{12/14}$-APG (Plantacare 1200) in a ratio of 2:1 active substance (Plantacare PS 10). Alkyl polyglycoside disrupts the formation highly viscous hexagonal phase of the alkyl ether sulfate which, in turn, inhibits the crystallization of alkyl polyglycosides (Fig. 41). The mixture is pumpable above 15°C and can be cold-processed with normal mixing units (for example Ekato propeller). On dilution, viscosity passes through a flattened broad maximum at 40% to 45% AS [69,80].

2. Formulation Technique

Foaming is an essential quality feature of cosmetic cleansing formulations. Alkyl polyglycosides foam considerably better than fatty alcohol ethoxylates, the foam volume increasing with increasing percentage of short carbon chains in the alkyl polyglycosides. They are comparable with betaines and sulfosuccinates but do not match the initial foam behavior or foam volume of alkyl (ether) sulfates, such as sodium lauryl ether sulfate (SLES; INCIname: Sodium Laureth Sulfate, sometimes referred to as fatty alcohol ether sulfate, FAES) [69,70,71,74,77] (Fig. 42). On the other hand, alkyl polyglycosides can stabilize the foam of anionics in hard water and in the presence of sebum so that up to 20% of total surfactant can be saved for the same foaming power [69,70,74].

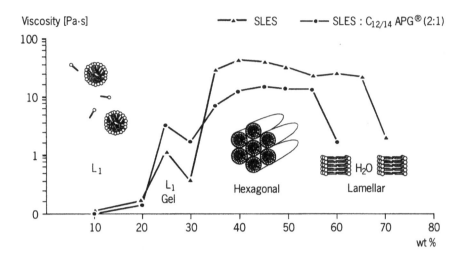

FIG. 41   Rheology of Sodium Laureth Sulfate (SLES) and Plantacare 1200 ($C_{12/14}$-APG) in admixture.

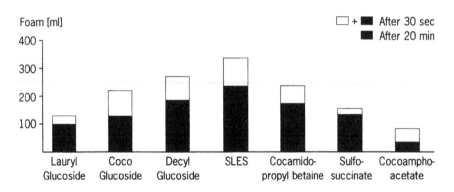

FIG. 42   Foaming properties of surfactants (1g AS/L, 15 °dH, 0.1 g/L sebum, perforated disk method DIN 53902).

Alkyl polyglycosides with alkyl chains longer than C_{12} contribute readily to the buildup of rodletlike mixed micelles in solutions of anionics and thus make a considerable contribution towards increasing viscosity [70,77] (Fig. 43). This effect on the one hand is somewhat weaker in standard ether sulfate formulations than with alkanolamides, but on the other hand is more pronounced with sulfosuccinates and highly ethoxylated alkyl ether sulfates which are very difficult to thicken with alkanolamides. Alkyl polyglycoside formulations without anionics

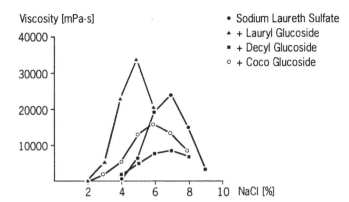

FIG. 43 Increase of viscosity by alkyl polyglycosides (10% AS SLES and 3% AS alkyl polyglycoside at 25°C).

can best be thickened by adding polymeric thickeners such as xanthan gum, alginate, polyethoxylated esters, carbomers, etc.

3. Effects on Hair

The mildness of alkyl polyglycosides toward the skin is also reflected in a caring effect on damaged hair. The tensile strengths of permed hair tresses are reduced far less by treatment with alkyl polyglycoside solutions than by standard ether sulfate solutions [70,74]. By virtue of these caring properties and their alkali stability, alkyl polyglycosides are also suitable as surfactants in coloring, permanent-wave, and bleaching formulations. Investigations of permanent-wave formulations revealed that the addition of alkyl polyglycosides favorably influences the alkali treatability of the hair and the waving effect [81].

The substantivity of surfactants to hair influences combability. Measurements of shampoo formulations on wet hair by objective methods (combing robot) and subjective methods (half-head test) showed that alkyl polyglycoside does not significantly reduce wet combability. However, a synergistic reduction in wet combability of about 50% was observed in the case of mixtures of alkyl polyglycosides with cationic polymers. By contrast, dry combability is considerably improved by alkyl polyglycosides. Increased interactions between the single hair fibers provide the hair with volume and manageability [70,74].

The increased interactions and film-forming properties also contribute towards styling effects. All-round bounce makes the hair appear vital and dynamically aesthetic. The bounce behavior of a hair curl can be determined in an automated test arrangement (Fig. 44) which investigates the torsional properties of hair fibers (bending modulus) and hair curls (stretch forces, attenuation, frequency, and amplitude of oscillation).

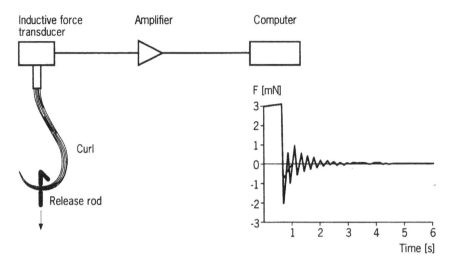

FIG. 44 Characterization of hair curl oscillations.

The hair curl is stretched by the release rod and the force function of the free attenuated oscillation is recorded by a measuring instrument (inductive force transducer) and processed by a computer. Styling products multiply the interactions of the individual hair fibers and lead to increased stretching work, amplitude, frequency, and attenuation values of the curl oscillation. Formulations of alkyl polyglycosides and protein hydrolyzate are as effective in regard to bounce and manageability as a 2% solution of polyvinyl pyrrolidone (PVP) [69], which is the conventional basis for hair spray formulations and setting lotions. From the practical point of view, easy rinsing of these products may also be an advantage over PVP formulations.

The effect of alkyl polyglycosides in setting lotions is further demonstrated by curl retention as compared with a water wave. Hair tresses on curlers are respectively treated with water and a setting lotion. After the treatment, the tresses are left hanging under their own weight in a chamber with controlled relative humidity (rH). The change in overall length after 6 hours is used as a measure of curl retention. The combination of Lauryl Glucoside and Hydrolyzed Collagen is equivalent to the PVP formulation with 75% curl retention at 70% rH [69].

In rinses and conditioners based on fatty alcohol and quaternary ammonium compounds, the synergism of alkyl polyglycoside/quaternary ammonium compound is favorable in reducing wet combability whereas dry combability is only slightly reduced in these applications. Oil components may also be incorporated in the formulations, further reducing the necessary quaternary content and imparting improved luster to the hair. Such o/w emulsions may be used as "rinse-off" or "leave-on" preparations for the aftertreatment of hair [69].

C. Miscellaneous Applications

By a special process based on brief exposure to high temperatures (flash drying), a water-containing paste of $C_{12/14}$-APG can be converted into a white noncaking alkyl polyglycoside powder with a residual moisture content of ~ 1% [82]. Alkyl polyglycosides may thus also be used in soaps and syndets. They exhibit good foam and skin feel properties and, by virtue of their excellent skin compatibility, represent an attractive alternative to conventional syndet formulations based on alkyl sulfates.

Similarly, $C_{12/14}$-APG may be used in toothpastes and other oral hygiene formulations. Alkyl polyglycoside/fatty-alcohol sulfate combinations show improved mildness toward oral mucous membrane and, at the same time, produce a rich foam. $C_{12/14}$-APG was found to be an effective booster for special antibacterial agents, such as chlorohexidine. In the presence of alkyl polyglycoside, the quantity of bactericidal agent can be reduced to about one-quarter without losing any bactericidal activity. This provides for the everyday use of high-activity products (mouth washes) which would otherwise be unacceptable to the consumer because of their bitter taste and their discoloring effect on the teeth [83].

D. Formulations

By virtue of their physicochemical and performance profile, alkyl polyglycosides are a class of products which represent a new concept in compatibility and care in cosmetics. Alkyl polyglycosides are multifunctional raw materials which are moving closer to the center of modern formulation techniques. They may advantageously be combined with conventional components and can even replace them in new types of formulations. To exploit the rich spectrum of supplementary effects of alkyl polyglycosides on the skin and hair, changes have to be made in conventional formulation systems involving the widely used alkyl (ether) sulfate/betaine combinations.

1. Shampoos

In a half-head test of two formulations, alkyl polyglycoside and betaine as primary surfactants were evaluated in regard to initial foam, foam volume, and feel. Although the results obtained were comparable, alkyl polyglycosides are preferred to betaines for their skin compatibility.

It is well-known that combinations of betaine and quaternaries only provide conditioning effects in the presence of anionic surfactants. Neither alkyl polyglycosides on their own nor the addition of betaine leads to any reduction in wet combability. However, in combination with cationic substances, alkyl polyglycosides significantly reduce wet combability with or without addition of betaines. To provide alkyl polyglycoside containing formulations with conditioning properties, any of the usual cationic materials may be used, including for example cationic proteins, cationic cellulose and guar derivatives, polyquaternium types, etc.

In addition to the use of alkyl polyglycosides in very mild formulations for frequent use, alkyl polyglycosides can be also used with advantage in special products for greasy hair and fine hair. In this case, a mild surfactant base is combined with special hair care ingredients which do not challenge the hair.

The development of specific cleansing concentrates (shampoo concentrates) as a new marketing concept is a consequence of consumer demand and legislation to reduce packaging waste. Highly concentrated raw materials are required for such formulations to control the water balance. However, formulations of this type are normally very viscous and require special processing techniques for dilution. By adding alkyl polyglycosides, it is possible to develop highly concentrated yet easily dilutable consumer products.

Wet hair is easy to comb while dry hair shows volume and lesser manageability. A special protein derivative, Potassium Abietoyl Hydrolyzed Collagen, serves as an effective ingredient against refatting of the hair. This was confirmed by objective measurements using the Shadow Imaging Method and by a salon test [84].

Another important speciality application are baby care shampoos which utilize the excellent dermatological compatibility of alkyl polyglycosides as a surfactant base in combination with other mild ingredients. For conditioning effects, incorporation of a cationic wheat protein hydrolyzate has a remarkably low irritation potential in the hens' egg test (HET-CAM).

2. Shower Baths

The feeling on the skin of shower baths and gels is an important factor in consumer acceptance. A pleasant feeling on the skin can be achieved with the same conditioning agents (cationic carbohydrate or protein derivatives) as described for shampoos.

An interesting concept for the highly regarded "shower and lotion" products is the incorporation of oil-containing emulsions in a surfactant base. The skin care effect of such formulations can be evaluated, for example, by determination of the transepidermal water loss (TEWL). Products are applied to the inside of the right and left forearms of volunteers. One hour after the treatment, the TEWL is measured with an evaporimeter. The TEWL value of oil-containing formulations is reduced compared to the formulation base.

Shower baths with peeling properties, a new market trend, can also be formulated with alkyl polglycosides. Soft abrasive particles, such as hardened jojoba wax, walnut bark, cellulose granules, apricot kernels, etc., are incorporated to remove scales and to achieve a specific feel. In addition, the abrasive particles promote circulation through a gentle massaging effect on the skin. The exfoliating effect must be carefully adapted to the type of skin and the specific consumer group. The products should exhibit good initial foam, a creamy feel, and a slight to medium exfoliating effect.

3. Body Wash Preparations

Good foaming and refatting performance characterize alkyl polyglycoside microemulsions as bath oils which are mild to the skin and which break up through dilution on application. Facial cleansers are speciality formulations. The dermatological properties of alkyl polyglycosides and their good deep pore-cleansing effects commend them for use in many types of clear formulations utilizing the phase behavior of medium- and long-chain alkyl polyglycosides.

4. Hair Treatments

For the same reasons and by virtue of their positive influence on combability, coupled with the fact that they may readily be combined with cationic compounds, alkyl polyglycosides can be formulated as the basis of rinses and conditioners. In combination with an esterquat as a conditioner, alkyl polyglycosides can be used to develop rinses based on vegetable materials. The effects of alkyl polyglycosides in styling formulations are based on their substantivity to hair. Styling gels and lotions are easily formulated using carbomers for viscosity adjustment. In permanent-wave formulations, the use of alkyl polyglycosides provides for an improved permanent-wave effect and, at the same time, reduces hair damage.

VI. SAFETY AND ENVIRONMENTAL ASPECTS

A. Toxicology of Alkyl Polyglycosides

Acquiring knowledge of the potential effects of substances on human beings and their environment is among the basic tenets in the worldwide "responsible care" initiative. To take responsible care seriously into account, manufacturers must ensure that their products do not endanger the health of consumers, the work force, or the general public. This responsibility relates both to direct exposure during production and use and to indirect exposure via the environment, including food and water.

All studies within this program were conducted in compliance with internationally accepted guidelines, especially the OECD Guidelines for the Testing of Chemicals and the Principles of Good Laboratory Practice [85].

1. Acute Toxicity

The acute toxicity of alkyl polyglycoside was comprehensively investigated. Variations of the test conditions involved homologues using either the short-chain ($C_{8/10}$) or the medium-chain ($C_{12/14}$) alkyl polyglycoside. Other modifications included the application route, which was either oral or dermal.

(a) Acute Toxicity After Ingestion. Tests [86–88] were performed in accordance with OECD guideline No. 401 and also followed the U.S. Toxic Substances Control Act (TSCA-40 CFR798) and the Federal Insecticide, Fungicide and Rodenticide Act (FIFRA-40 CFR158, 162). The results of the tests are set out in Table 16. On the basis of these tests, there would be no risk from alkyl polyglycosides if swallowed. Under E.U. and U.S. classification rules, alkyl polyglycosides do not require hazard classification or labeling.

(b) Acute Toxicity After Contact with the Skin. Tests [89,90] were carried out under OECD guideline No. 402 and also under the TSCA and FIFRA regulations. The results are shown in Table 17. On the basis of these data, contact with

TABLE 16 Alkyl Polyglycosides ($C_{8/10}$-, $C_{12/14}$APG) Tested on Acute Oral Toxicity in Rats

Chain length	$C_{8/10}$	$C_{12/14}$	$C_{12/14}$
Degree of polymerization	1.6	1.6	1.4
Concentration (% active substance)	50	50	60
Rat strain	Sprague-Dawley	Sprague-Dawley	Wistar
Number of animals	5 - 5	5 - 5	2 - 2
Sex, m = male; f = female	m - f	m - f	m - f
Mortality	0/5 - 0/5	0/5 - 0/5	0/2 - 0/2
Animals with gross necropsy observations	0/5 - 0/5	0/5 - 0/5	0/2 - 0/2
LD_{50} (mg/kg body weight)	>5000	>5000	>2000
Reference	[2]	[3]	[4]

TABLE 17 Alkyl Polyglycosides ($C_{8/10}$-, $C_{12/14}$ APG) Tested on Acute Dermal Toxicity in Rabbits

Chain length	$C_{8/10}$	$C_{12/14}$
Degree of polymerization	1.6	1.6
Concentration (% active substance)	50	50
Rabbit strain	New Zealand White	New Zealand White
Number of animals	5–5	5–5
Sex, m = male; f = female	m–f	m–f
Mortality	0/5–1/5[a]	0/5–0/5
LD_{50} (mg/kg body weight)	>2000	>2000
Reference	[5]	[6]

[a]One mortality due to Tyzzer's disease.

the skin is unlikely to involve any risk. Alkyl polyglycoside is not acute toxic after dermal application under the E.U. classification scheme.

2. Dermal Irritation

Investigations were conducted in accordance with OECD guideline No. 404, the international standard method for testing dermal irritation. The results are shown in Table 18. The data provide clear information on structure-response and concentration-response relationships. Short-chain ($C_{8/10}$) alkyl polyglycoside in commercial concentrations would appear to have no irritating effects [91,92].

Medium-chain ($C_{12/14}$) alkyl polyglycosides in the same concentration range (40% to 60%) are irritating to the skin [93–95]. The irritation responses justifying classification start at concentrations exceeding 30%; i.e., concentrations > 30% cannot be classified as "irritating to the skin" [96,97]. With increasing concentration, the number of animals responding with erythema increases. At 60%, a plateau is reached. Concentrations of 60% and higher produced significant erythema in all the test animals. The same applies to animals that had received a 100% alkyl polyglycoside sample: a $C_{12/14}$ alkyl polyglycoside in a concentration of 100% was irritating, but not corrosive [98].

Dermal irritation was not pH dependent (data not shown in the table). A sample with a pH of 7 produced responses similar to those of the commercial product with a pH of 11.5 [99,100].

From the results of the skin irritation tests, it can be concluded that $C_{8/10}$ alkyl polyglycoside and $C_{12/14}$ alkyl polyglycoside in a concentration of up to 30% do not require classification or labeling. $C_{12/14}$ alkyl polyglycosides in concentrations of >30% to 100% fall within the R38 ("irritating to the skin") category of the E.U. risk classification.

TABLE 18 Alkyl Polyglycosides ($C_{8/10}$-, $C_{12/14}$ APG) Tested on Dermal Irritation in Rabbits

Chain length	Concentration (% active substance)	PD II	"Positive" responder = animals with score >2			Mean values (24/48/72 h)		Classification as "irritating"	Ref.
			Erythema			Erythema	Edema		
			Abs.	In %	Edema				
$C_{8/10}$	50	1.30	0/3	0	0/3	1.1	0.3	no	[91]
	60	0.80	0/3	0	0/3	0.9	0.0	no	[92]
$C_{12/14}$	50	1.27	2/3	67	0/3	1.8	1.8	yes, R38	[93]
$C_{12/14}$	20	1.20	0/3	0	0/3	1.2	1.25	no	[94]
	30	2.80	1/6	17	0/6	1.7	0.9	no	[95]
	40	1.50	2/3	67	0/3	2.1	0.9	yes, R38	[96]
	60	1.99	4/4	100	3/4	2.9	2.1	yes, R38	[97]
	100	3.70	3/3	100	0/3	2.2	1.6	yes, R38	[98]

3. Mucous Membrane Irritation (Eye Irritation)

In vivo irritation in rabbits' eyes were conducted in accordance with OECD guideline No. 405 under which the substance to be tested is applied to one eye of each of at least three rabbits. The untreated eye of each animal serves as control. An aliquot of a 0.1-mL aqueous solution of the sample was instilled once into the eyes of four rabbits for an intended contact time of 24 hours.

The eyes were assessed by awarding scores at certain intervals after application under the Draize scheme. The 24/48/72-hour mean scores were determined for the cornea, for conjunctiva, and for the iris. All responses were checked for reversibility for 21 days. The results are given in Table 19. $C_{8/10}$ alkyl polyglycoside does not have to be classified as a hazard or labeled as such [101], whereas $C_{12/14}$ alkyl polyglycosides fall within the R36 ("irritating to the eyes") category of the E.U. risk classification [102].

4. Skin Sensitization

The international standard test method is OECD guideline No. 406 (corresponding to E.U. guideline No. B6), which also lists the test protocols generally accepted by the relevant authorities. Tests were carried out either by the Magnusson/Kligman method or by the Buehler method to investigate the sensitizing potential of a $C_{12/14}$ alkyl polyglycoside in female guinea pigs of the Pirbright White strain. The only difference between the two methods lies in the mode of induction. The test conditions and results are summarized in Table 20.

None of the tests produced responses which could be attributed to an allergic reaction [103–106]. On the evidence of these tests, alkyl polyglycoside does not require classification or labeling. This was confirmed by a human repeated insult patch test in which alkyl polyglycoside did not induce any sensitization in volunteers. Thus, the animal model provided clear predictions of effects on human beings.

5. Mutagenicity

(a) Gene Mutations. Gene mutations can be investigated by the *Salmonella typhimurium* reversion (Ames) test. The international standard test method is OECD guideline No. 471. This microbial assay is based on reverse mutations of

TABLE 19 Alkyl Polyglycosides ($C_{8/10}$-, $C_{12/14}$ APG) Tested on Eye Irritation in Rabbits

| Chain length | Concentration (% AS) | Mean values (24/48/72 h) | | | Classification as "irritating" | Reference |
		Cornea	Iris	Conjunctiva (erythema)		
$C_{8/10}$	70	1.8	0.6	2.4	no	[101]
$C_{12/14}$	50	1.25	0.7	2.5	yes, R36	[102]

TABLE 20 Alkyl Polyglycosides ($C_{8/10}$-, $C_{12/14}$ APG) Tested on Skin Sensitization in Guinea Pigs

Method	Concentrations (% AS) for Inductions	Challenge	Number of positive responders	Classification as "sensitizing to the skin"	Reference
Magnusson-Kligman	Intracutaneous: 1% Topical: 60%	10%	0/20	no	[103]
	Intracutaneous: 0.1%	1.25	0/20	no	[104]
	Topical: 10%	2.5	0/20		
Buehler	Topical: 20%	20%	0/20	no	[105]
	Topical: 12.5%	9%	0/20	no	[106]

Salmonella typhimurium from auxotrophism histidine-dependent) to prototrophism (non-histidine-dependent). When mutated *Salmonella typhimurium* is exposed to a mutagen, mutation to the non-histidine-dependent form takes place in a proportion of the bacterial population.

Alkyl polyglycoside did not induce reverse mutations in the tested strains of *Salmonella typhimurium* either with or without metabolic activation. Accordingly, alkyl polyglycoside is regarded as nonmutagenic in this in vitro bacterial mutagenicity test [107].

(b) Chromosomal Mutations. Chromosome mutations can be detected by the in vitro cytogenetic test in Chinese hamster V79 lung fibroblasts. The international standard test method is OECD guideline No. 473. This in vitro mammalian cytogenetic test indicates the damage to chromosomes by structural aberrations or may provide an indication of the induction of numerical aberrations by chemical substances. No biological activity in the induction of chromosomal aberrations was observed after application of alkyl polyglycoside. There was no significant increase in chromosomal aberrations after the treatment by comparison with the control values. Accordingly, alkyl polyglycoside is considered to be nonmutagenic in this chromosome aberration test [108].

6. Toxicokinetics and Metabolism

Toxicokinetics is the study of the rates of absorption, distribution, metabolism, and excretion of substances. Metabolism describes all the processes by which a particular substance is handled in the human body. Literature studies are available on the behavior of several alkyl—glycosides in the mammalian organism. Octyl—D-[U-^{14}C]-glucoside, [1-^{14}C]-dodecyl—D-maltoside, and [1-^{14}C]-hexadecyl—D-glucoside were used as test substances representative of alkyl polyglycosides [109]. The toxicokinetic studies demonstrated that alkyl polyglycosides are physiologically compatible.

7. Subchronic Toxicity

Subchronic toxicity is the adverse effect produced by the repeated daily administration of a chemical substance to experimental animals for part (not more than 10%) of their lifespan. The international standard test method for oral application is OECD guideline No. 408. A subchronic toxicity study was carried out with a $C_{12/14}$ alkyl polyglycoside in male and female Wistar rats over a period of 90 days.

According to the results [110], a daily dose of 1000 mg/kg did not lead to any cumulatively toxic effects. This dose is thus classified as a "no-observed-adverse-effect level" (NOAEL).

8. Conclusions

The object of toxicological studies is to assess the hazard potential of chemical products in order to asses the risks to the health of the user and to prevent possible adverse effects, even after improper use. The biological endpoints which have to be considered in toxicological studies follow the state of science and, for the most part, are included in current legal regulations (EC Directive 93/35/EEC).

Alkyl polyglycosides varying in chain length and purity have been subjected to the most relevant toxicological endpoints necessary for a sound risk assessment. On the basis of the information available, alkyl polyglycosides are not considered as toxic or harmful in acute toxicity tests, but in high concentrations have to be classified as irritating to the skin and eyes. In addition, sensitizing effects are unlikely to occur. A NOAEL of 1000 mg/kg body weight can be estimated for toxicity after repeated oral application. In in vitro tests, alkyl polyglycosides did not show any potential for gene or chromosome mutations.

Safety in use and handling also includes foreseeable improper use, such as the accidental swallowing of cosmetic products by children. By virtue of their low acute oral toxicity, alkyl polyglycosides do not contribute to the toxicity of cosmetic and other household products. The acute oral toxicity values (LD_{50}) are of the order of several grams per kilogram body weight. In other words, it is virtually impossible to be seriously poisoned by these products.

B. Ecological Properties of Alkyl Polyglycosides

Alkyl polyglycosides have been extensively investigated with regard to their fate and effect in the environment [111,112]. The prescribed OECD method for detecting the biological primary degradation of nonionics is not applicable to alkyl polyglycosides because they do not contain any ethylene oxide groups and are thus not BiAS-active. Nevertheless, it can be extrapolated from the very favorable ultimate degradation data that the primary degradation step also proceeds with ease. This was confirmed in the OECD Confirmatory Test by applying an

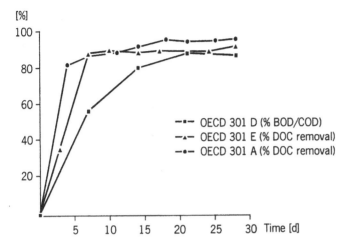

FIG. 45 Biodegradation kinetics of $C_{12/14}$-APG in standard tests.

alkyl polyglycoside–specific analysis method. Ready ultimate biological degradation was observed with complete mineralization and/or assimilation of alkyl polyglycosides under both aerobic and anaerobic conditions.

In the Closed Bottle Test (OECD 301), the "ready biodegradability" criterion under aerobic conditions is fulfilled with the "10-day-window" requirement under which 60% degradation must occur within 10 days of passing the 10% degradation level (Fig. 45).

The ultimate degradation of alkyl polyglycosides without residues and stable metabolites was confirmed in a modified Coupled-Units Test, the so-called Test for Recalcitrant Metabolites. In this test, the effluent of the test unit is circulated for about 6 weeks to detect possible accumulations of nonreadily degradable substances [112].

Under European standards, an overall evaluation of the environental risk of alkyl polyglycosides requires a realistic description of a scenario in which assessment of exposure in the environment of a substance (predicted environmental concentration) is compared with its effect in the environment (predicted no-effect concentration) [113]. Such an evaluation leads to the conclusion that, even under unfavorable conditions, no environmental risk is involved in the use of alkyl polyglycosides.

The ecotoxicological effects which, given rapid biodegradability of the substances, continue to lose relevance in an overall ecological risk assessment, show favorable findings in all test systems. The no-observed-effect concentrations (NOEC) for acute or chronic toxicity, as determined on single-species or biocenotic communities of the aquatic and terrestrial environment, show that alkyl polyglycosides have comparatively low ecotoxicity [111].

REFERENCES

1. E Fisher. Ber 26:2400, 1893.
2. B Capon. Chem Rev 69:389, 1969, and RJ Ferrier. Fortschr Chem Forsch 14:389, 1970.
3. E Fischer, L Beensch. Ber 27:2478, 1894, and E Fischer. Ber 28:1145, 1895.
4. W Koenigs, E Knorr. Ber 34:957, 1901.
5. E Fischer, B Helferich. Justus Liebigs Ann Chem 386:68, 1911.
6. P Schulz. Chim Oggi 10:33, 1992.
7. K Toshima, K Tatsuta. Chem Rev 93:1503, 1993, and R Schmidt in E Winterfeldt, ed. Comprehensive Organic Synthesis. Oxford: Pergamon Press, 1991, Vol 6, p 33.
8. DRP 593422, H. Th. Böhme AG (1934) and DRP 611055, H. Th. Böhme AG (1935).
9. J Knaut, G Kreienfeld. Chim Oggi 11:4, 1993.
10. PT Schulz. Proc. BACS Symp., Chemspec Eur. 91, Amsterdam, 1991, p 33, and PT Schulz. Chim Oggi 10:33, 1992.
11. M Biermann, K Schmid, PT Schulz. Starch/Stärke 45:281, 1993.
12. K Hill, M Weuthen. Spektrum der Wissenschaft, June:113, 1994.
13. PM McCurry. Henkel Corporation, unpublished results.
14. PJ Flory. J Am Chem Soc 74:2718, 1952.
15. EP 0437460 B1, Henkel (1988).
16. EP 0495174, Hüls (1991).
17. EP 0617045 A2, Akzo (1994).
18. EP 0448799, Hülls (1990).
19. WP 94/04544, BASF (1992).
20. EP 0357969 B1, Henkel (1988).
21. EP 0301298 B1, Henkel (1987).
22. EP 0482325, Hüls (1990).
23. W Ruback, S Schmidt. In H van Bekkum, ed. Carbohydrates as Organic Raw Materials III. Weinhem: VCH Verlagsgesellschaft, 1996, p 231, and EP 0514627, Hüls (1991).
24. EP 0099183, Staley (1982).
25. WO 93/10133, Henkel (1991).
26. YZ Lai, F Shafizadeh. Carbohydr Res 38:177, 1970.
27. GR Ponder, GN Richards. Carbohydr Res 208:93, 1974.
28. WO 90/06933, Henkel (1990).
29. EP 0492397, Kao (1990).
30. DE-OS 3833780, Henkel (1988).
31. DE-OS 3932173, Henkel (1989).
32. H Waldhoff, J Scherler, M Schmitt. Proceedings 4th World Surfactant Congress, Barcelona, June 1996, Vol 1, p 507.
33. R Spilker, B Menzebach, U Schneider, I Venn. Tenside Surf Det 33:21, 1996.

34. N Buschmann, A Kuse, S Wodarczak. Agro Food Industry, Hi-Tech, Jan/Feb:6, 1996.
35. N Buschmann, S Wodarczak. Tenside Surf Det 32:336, 1995.
36. A Bruns, H Waldhoff, W Winkle. Chromatographia 27:340, 1989.
37. Henkel Corp., unpublished results.
38. K Shinoda, T Yamaguchi, R Hori. Bull Chem Soc Jpn 34:237, 1961.
39. T Böcker, J Thiem. Tenside Surf Det 26:318, 1989.
40. D Nickel, C Nitsch, C-P Kurzendörfer, W von Rybinski. Prog Colloid Polym Sci 89:249, 1992.
41. F Jost, H Leiter, MJ Schwuger. Colloid Polym Sci 266:554, 1988.
42. T Förster, H Hensen, R Hofmann, B Salka. Cosmet Toiletries 110:23, 1995; Parfumerie Kosmetik 76:763, 1995.
43. D Nickel, C Nitsch, C-P Kurzendörfer, W von Rybinski. Prog Colloid Polym Sci 89:249, 1992.
44. EM Kutschmann, GH Findenegg, D Nickel, W von Rybinski. Colloid Polym Sci 273:565, 1995.
45. K Fukuda, O Södermann, B Lindman, K Shinoda. Langmuir 9:2921, 1993.
46. T Förster, B Guckenbiehl, A Ansmann, H Hensen. Seifen Ole Fette Wachse J 122:746, 1996.
47. M Weuthen, R Kawa, K Hill, A Ansmann. Fat Sci Technol 97:209, 1995.
48. CF Putnik, NF Borys. Soap Cosmet Chem Spec 86:34, 1986, and FA Hughes, BL Lew, J Am Oil Chem Soc 47:162, 1970.
49. H Andree, B Middelhauve. Tenside Surf Det 28:413, 1991, and PA Siracusa. HAPPI April 92:100, 1992.
50. C Nieendick, K Schmid. Seifen Ole Fette Wachse J 121:412, 1995.
51. K Schmid. 6e Giornale CID Congress, Rome, 1995.
52. K Schmid. In H Eierdanz, ed. Perspektiven nachwachsender Rohstoffe in der Chemie. Weinheim: VCH Verlagsgesellschaft, 1996, p 41, and P Jürges, A Turowski. In H Eierdanz, ed. Perspektiven nachwachsender Rohstoffe in der Chemie. Weinheim: VCH Verlagsgesellschaft, 1996, p 61.
53. B Jackwerth, H-U Krächter, W Matthies. Parfumerie Kosmet 74:142, 1993.
54. W Sterzel, FG Bartnik, W Matthies, W Kästner, K Künstler. Toxicol In Vitro 4:698, 1990, and H Spielmann, S Kalweit, M Liebsch, T Wirnsberger, I Gerner, E Bertram Neis, K Krauser, R Kreiling, HG Miltenburger, W Pape, W Steiling. Toxicol In Vitro 7:505, 1993.
55. H Andree, B Middelhauve. World Surfactant Congress, Montreux, Switzerland, 1992.
56. W Matthies. Seifen Ole Fette Wachse J 119:922, 1993.
57. F Hessel. In K Hill, W von Rybinski, G Stoll, eds. Alkyl Polyglcosides. Weinheim: VCH Verlagsgesellschaft, 1997, pp 111–115.
58. H Heitland, H Marsen. In J Falbe, ed. Surfactants in Consumer Products New York: Springer-Verlag, 1987, p 306.
59. B-D Holdt, P Jeschke, R Menke, JD Soldanski. Seifen Ole Fette Wachse J 120:42, 1994.
60. IPP Quality Standard. Seifen Ole Fette Wachse J 112:371, 1986.
61. DIN 53449, Parts 1–3. Evaluation of Environmental Stress Cracking.

62. H Upadek, P Krings. Seifen Ole Fette Wachse J 117:554, 1991.
63. DE 3920480, Henkel (1989).
64. R Puchta, P Krings, H-M Wilsberg. Seifen Ole Fette Wachse J 116:241, 1990.
65. D Balzer, N Ripke. Seifen Ole Fette Wachse J 118:894, 1992.
66. EP 553099, Henkel (1991).
67. F Hirsinger, KP Schick. Tenside Surf Det 32:193, 1995.
68. W Matthies, H-U Krächter, W Steiling, M Weuthen. 18th IFSCC, Venice, 1994, Poster Vol 4, p 317.
69. P Busch, H Hensen, J Kahre, H Tesmann. Agro-Food-Ind Hi-Tech (Sept/Oct): 23, 1994.
70. P Busch, H Hensen, H-U Krächter, H Tesmann. Cosmet Toiletries Manuf Worldwide 123, 1994.
71. KH Schrader, M Rohr. Eur Cosmet 18, 1994.
72. KH Schrader. Parfumerie Kosmet 75:80, 1994.
73. Henkel KGaA (1995), unpublished results, Rep. No. R9500783.
74. P Busch, H Hensen, H Tesmann. Tenside Surf Det 30:11675, 1993.
75. B Jackwerth, H-U Krächter, W Matthies. Parfumerie Kosmet 74:143, 1993 (Engl. ed 74:142, 1993).
76. DD Strube, SW Koontz, RI Murata, RF Theiler. J Soc Cosmet Chem 40:297, 1989.
77. B Salka. Cosmet Toiletries 108:89 1993.
78. M Rohr, KH Schrader. Eur Cosmet 8:24, 1994.
79. U Zeidler. J Soc Cosmet Chem Jpn 20:17, 1986.
80. T Förster, H Hensen, R Hofmann, B Salka. Cosmet Toiletries 110:23, 1995; Parfumerie Kosmet 76:763, 1995.
81. J Kahre, D Goebels. Agro-Food-Ind Hi-Tech (March/April: 29, 1995; 41st Annual Conference SEPAWA 1994, p 36.
82. DE-P 19534371, Henkel (1995).
83. EP 0304627 A2, Henkel (1988).
84. P Busch, H Hensen, D Fischer, A Ruhnke, J Franklin. Seifen Ole Fette Wachse J 120:339, 1994; Cosmet Toiletries 110:59, 1995.
85. Organization for Economic Cooperation and Development—OECD (1981), Paris.
86. Henkel Corp. (1987), unpublished results, Rep. No. TBD EX 0321.
87. Henkel Corp. (1990), unpublished results, Rep. No. R9600989.
88. Henkel KGaA (1986), unpublished results, Rep. No. TBD860297.
89. Henkel Corp. (1987), unpublished results, Rep. No. TBD EX 0323.
90. Henkel Corp. (1990), unpublished results, Rep. No. R9600990.
91. Henkel KGaA (1993), unpublished results, Rep. No. R9300408.
92. Henkel KGaA (1993), unpublished results, Rep. No. R9300407.
93. Henkel KGaA (1993), unpublished results, Rep. No. R9400 116.
94. Henkel KGaA (1988), unpublished results, Rep. No. TBD880089.
95. Henkel KGaA (1993), unpublished results, Rep. No. R9300115.
96. Henkel KGaA (1988), unpublished results, Rep. No. TBD880405.
97. Henkel KGaA (1994), unpublished results, Rep. No. R9400725.
98. Henkel KGaA (1994), unpublished results, Rep. No. R9400459.
99. Henkel KGaA (1993), unpublished results, Rep. No. RT930139.

100. Henkel KGaA (1993), unpublished results, Rep. No. R930138.
101. Henkel Corp. (1990), unpublished results, Rep. No. R9601003.
102. Henkel Corp. (1990), unpublished results, Rep. No. R9601004.
103. Henkel KGaA (1990), unpublished results, Rep. No. TBD900290.
104. Henkel KGaA (1988), unpublished results, Rep. No. TBD880412.
105. Henkel KGaA (1994), unpublished results, Rep. No. R9400208.
106. Henkel KGaA (1993), unpublished results, Rep. No. R9300256.
107. Henkel KGaA (1990), unpublished results, Rep. No. 900467.
108. Henkel KGaA (1995), unpublished results, Rep. No. 9400243.
109. N Weber. Fette Seifen Anstrichmittel 86:585, 1984.
110. Henkel KGaA (1989), unpublished results, Rep. No. 890191, and W Sterzel. In C Gloxhuber, K Künstler, eds. Anionic Surfactants: Biochemistry, Toxicology, Dermatology. Surfactant Science Series 43. New York: Marcel Dekker, 1992, p 411.
111. J Steber, W Guhl, N Stelter, FR Schröder. Tenside Surf Det 32:515 1995.
112. P Gerike, W Holtmann, W Jasiak. Chemosphere 13:121, 1984.
113. EEC (1994) Commission Regulation (EC) No. 1488/94; EEC (1994) Risk Assessment of Existing Substances, Techn. Guidance Document, European Commission DG XI, Brussels.

2

Nitrogen-Containing Specialty Surfactants

SHOAIB ARIF and FLOYD E. FRIEDLI
Goldschmidt Chemical Corporation, Dublin, Ohio

I. INTRODUCTION

The major use of nitrogen derivatives in laundry is as fabric softeners, where quaternary ammonium salts are used (see Chapter 9). However, ethoxylated amines, various types of amides, amphoterics/betaines, amine oxides, and other analogs are quite useful in the laundry detergent area [1–3], which is one focus of this chapter. Other key topics are detergency relating to cleaning of various hard surfaces, carpets, and hair.

The current trend for household laundry is concentrated detergents, both liquids and powders, where packages are smaller, lighter, more convenient, and shipping is less costly. There is a strong need to identify new surfactants that could offer one or more of the following advantages:

Better detergency per gram of surfactant so concentration can be even higher
Good detergency on very difficult stains
Coupling or formulation properties
Antiredeposition, dye transfer inhibition, resistance to hard water, enzyme stability, or other added benefits
Excellent safety and environmental profile

These properties will be discussed as they apply to nitrogen-based surfactants.

Laundry detergents are generally composed of linear alkylbenzene sulfonates (LAS) and alcohol ethoxylates (AE) plus enzymes, builders, and possibly a bleach. The major surfactants are well known for their particulate- and oil-removing properties, respectively, plus low cost and safety.

Ether sulfates (SLES) have long been known as excellent detergents with good hard-water tolerance. Alpha olefin sulfonates (AOS) are widely used in Japan and less used elsewhere. Recent studies have shown alpha-sulfomethyl esters as workable detergents, as are alkyl polyglycosides and alkyl glu-coseamides. These last two surfactants are presented in other chapters of this book.

Primarily owing to cost or even tradition, other surfactants, such as ethoxy-lated amines, amides, and amphoterics, have not been used generally in laundry detergents. However, these specialty surfactants have been included in a number of other household cleaners, industrial formulations, and personal-care products.

Nitrogen-containing surfactants like fatty amine derivatives and amphoteric surfactants demonstrate some unique performance properties in a wide variety of detergent formulations. The nitrogen develops a positive charge, particularly in acidic pH which provides the surfactant molecule some interesting properties. In combination with anionics, for example, they form mixed micelles which have relatively closer packing due to the neutralization of charges. Some of the most commonly used nitrogen-containing surfactants that are used in detergent indus-try are described in the remainder of this chapter.

II. ETHOXYLATED AMINES

A. Structure and Method of Preparation

Alkyl primary amines react with ethylene oxide (EO) to form a special type of nonionic surfactant (Fig. 1) [4–6]. Addition of the first 2 moles of EO occurs without catalyst, while more chain building requires a basic catalyst similar to ethoxylating an alcohol. To function as nonionics, R typically contains from 10 to 22 carbon while 5 to 20 moles $(x + y)$ of ethylene oxide is needed to give an

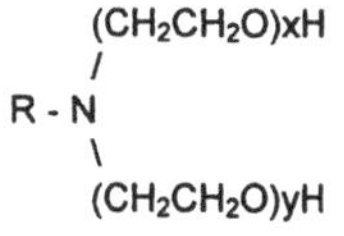

$$
\begin{array}{c}
(CH_2CH_2O)xH \\
/ \\
R\text{-}N \\
\backslash \\
(CH_2CH_2O)yH
\end{array}
$$

FIG. 1 Ethoxylated amine.

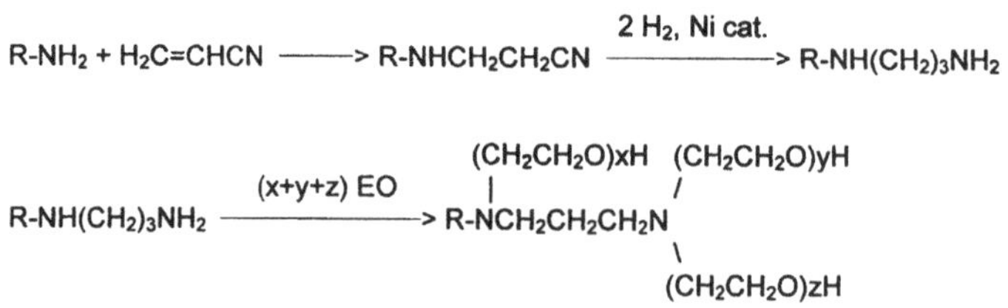

$$
R\text{-}NH_2 + H_2C{=}CHCN \xrightarrow{} R\text{-}NHCH_2CH_2CN \xrightarrow{2\,H_2,\ Ni\ cat.} R\text{-}NH(CH_2)_3NH_2
$$

$$
R\text{-}NH(CH_2)_3NH_2 \xrightarrow{\ (x+y+z)\ EO\ }
\begin{array}{c}
(CH_2CH_2O)xH \quad (CH_2CH_2O)yH \\
| \qquad\qquad\quad / \\
R\text{-}NCH_2CH_2CH_2N \\
\backslash \\
(CH_2CH_2O)zH
\end{array}
$$

FIG. 2 Ethoxylated polyamine.

appropriate HLB. In practice, the usual starting materials are coco or tallow amines. Propylene oxide can also be used, but is less common.

Primary amines can be reacted [5,7] with acrylonitrile to form an aminonitrile, which can be hydrogenated to a diamine. The sequence can be repeated if desired and these polyamines can be alkoxylated [4] to form another class of nonionics, polyalkoxylated polyamines (Fig. 2).

B. Detergency and Other Useful Properties

1. Detergency in Laundry

Fatty amine ethoxylates contain two ethylene oxide chains and a nitrogen atom, making the molecule basic. These two attributes make them different than the ordinary nonionic surfactants like alcohol ethoxylates (AE) and nonylphenol ethoxylates (NPE). Amine ethoxylates are amber-colored liquids that are almost 100% active. These low-foam nonionic surfactants have a natural pH of 9 to 10 and are good surface tension reducers. A 10% water solution of cocoamine with 5 moles moles EO gives a surface tension of 25 to 28 dynes/cm. The dynamic surface tension of cocoamine with 5 moles EO at various bubble frequencies is similar to C_{12-15} alcohol with 7 moles EO (this is the standard alcohol ethoxylate used for comparisons in this chapter, abbreviated AE). The critical micelle concentration (CMC) of the two surfactants is also similar (see Figs. 3 and 4). In laundry applications, ethoxylated fatty amines perform differently from the ordinary nonionics. AEs clean best at their cloud point, while ethoxylated amines, which do not have a cloud point, clean well at a wide variety of temperatures. Some other advantages of the ethoxylated fatty amines are:

Cold-water detergency
Stain removal
Provision of alkalinity
Antiredeposition of soil
Corrosion inhibition
No gelling
Low freeze point

Detergency performance of amine ethoxylates depends on the hydrophobe and the number of moles ethylene oxides attached. As the moles of ethylene oxide are added to the fatty group the detergency [8] increases to a certain point after which it does not change much (Table 1) [Tables 1–71 can be found beginning on page 76].

In general, the cocoamine with 5 moles EO and tallowamine with 10 moles EO have shown the best detergency performance. The detergency generally increases as the dosage of the surfactant increases. A dosage curve ran using tallowamine with 15 moles EO shows that the detergency increases until a certain

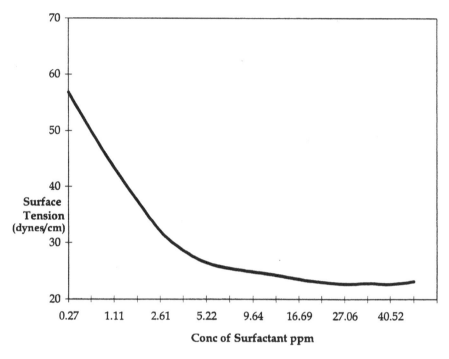

FIG. 3   CMC of C_{12-15} alcohol + 7 EO.

concentration of the surfactant in water is reached and then the curve flattens out. The results are shown in Figure 5. The hardness of wash water does not affect the detergency performance of amine ethoxylates to a great degree. The performance does go down but the drop is minor. Using the ASTM D-3050 procedure [8], the percent soil removal is shown in Table 2 for tallowamine with 15 moles EO.

One of the unique performance attribute of amine ethoxylates is their ability to remove grass stains. Amine ethoxylates show better grass stain removal than most anionic and nonionic surfactants used in laundry detergents (Table 3).

Amine ethoxylates show better soil removal performance than AE in some areas whereas AE are better in other areas. A comparison of detergencies of an amine ethoxylate and an AE is given in Table 4.

Most detergent formulas use a combination of various surfactants to balance their performance. Generally, one or more anionic and one or more nonionic surfactants are used in a formula. Amine ethoxylates work well with other surfactants. They are compatible with nonionic, anionic, and amphoteric surfactants. Unlike AE and NPE, amine ethoxylates do not normally form water/surfactant

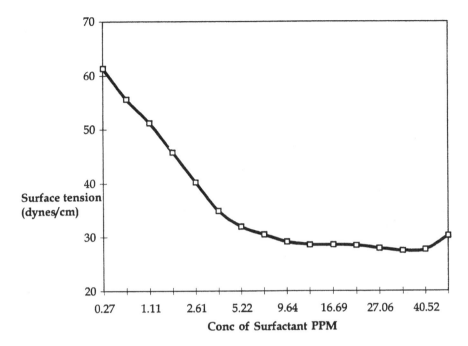

FIG. 4 CMC of cocoamine + 5 EO.

gels. High active detergents can be formulated using amine ethoxylates. Amine ethoxylates also do not exhibit a cloud point, so their cleaning is good over a broad range of temperature. Table 5 gives a comparison of the percent soil removal data generated by using blends of anionic surfactants with AE versus amine ethoxylates.

In a study, binary mixtures of five surfactants were evaluated for their physical properties and detergencies. The physical property data shows some trends, but does not correspond to the detergencies in a well defined matrix (Table 6). Lower dynamic surface tension usually gives better cleaning on dust sebum.

2. Dye Transfer Inhibition

One of the biggest consumer complaints about laundry detergent is the transfer of dyes from one fabric to another or at a different spot on the same fabric. This is called dye transfer. Polymers like Polyvinyl Pyrrolidone (PVP) are used in detergents to inhibit the transfer of dyes. Ethoxylated amines also show good dye transfer inhibition properties [9]. However, neither PVP nor amine ethoxylates amines work well as dye transfer inhibitors in presence of LAS. The reason probably is that LAS acts as a strong dye transfer or dye migrating agent. Both

TABLE 1 Detergency of Tallow Amine Ethoxylates at Different Levels of EO

Surfactant	Fabric	Soil	SRI	% SR
Tallowamine + 2 EO	Cotton	Dust/sebum	70.89	−14.30
	Cot/Pol		69.77	−14.08
Tallowamine + 6 EO	Cotton		91.01	65.97
	Cot/Pol		87.33	51.83
Tallowamine + 8 EO	Cotton		93.91	77.89
	Cot/Pol		91.19	66.40
Tallowamine + 10 EO	Cotton		94.66	80.92
	Cot/Pol		92.71	72.18
Tallowamine + 12 EO	Cotton		95.04	82.74
	Cot/Pol		92.93	73.03
Tallowamine + 14 EO	Cotton		95.06	82.51
	Cot/Pol		93.20	73.97
Tallowamine + 16 EO	Cotton		95.03	82.74
	Cot/Pol		93.20	73.99
Tallowamine + 18 EO	Cotton		94.99	82.51
	Cot/Pol		93.31	74.50
Tallowamine + 20 EO	Cotton		95.04	82.88
	Cot/Pol		93.30	74.42

SRI = soil removal index; SR = soil removed, Cot/Pol = cotton/polyester blend 1.6 g active surfactant per liter wash water; 150 ppm water hardness.

LAS as well as the disulfonates like alkyl diphenyloxide disulfonates (Chapter 4) are used as migrating and leveling agents during the dyeing process of clothe and carpet. In alkyl ether sulfates like SLES-based laundry detergent formulas, PVP has shown better performance over amine ethoxylates whereas in nonionic based formulas amine ethoxylates outperform PVP.

Monoamine ethoxylates having one nitrogen show good dye transfer inhibition properties when used at relatively high concentrations particularly as compared to PVP. From 1% to 2% of PVP does a satisfactory job of dye transfer inhibition in a liquid laundry detergent formula whereas 5% to 10% a monoamine ethoxylate may be required. The substitution of PVP with amine ethoxylates may still be cost-effective since PVP is about three to five times more expensive than amine ethoxylates, and the later act as a surfactant as well. In other words, one can take out 10% of alcohol ethoxylates from the formula and add 10% amine ethoxylate. The dye transfer inhibition properties of a 5-mole EO adduct of coconut amine were evaluated according to the ASTM procedure [10]. The results are listed in Table 7. This table gives a dosage curve for the dye transfer inhibition of increasing amounts of cocoamine ethoxylate as well as the binary surfactant systems based on amine ethoxylates and another

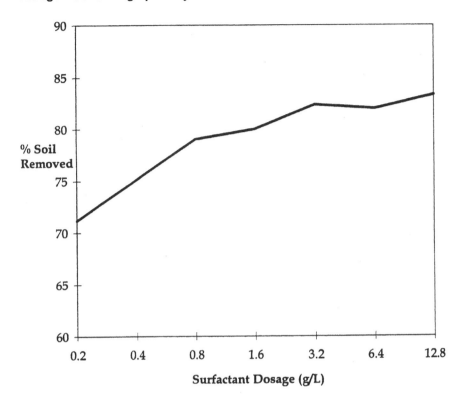

FIG. 5 Detergency curve: tallow amine + 15 moles EO.

TABLE 2 Effect of Hard Water on Detergency of Tallow Amine + 15 EO (100°F dust/sebum on cotton)

Water hardness	% Soil removed
50 ppm	80.0
150 ppm	80.6
300 ppm	78.9

commonly used surfactant like LAS, SLES, and AE. It can be seen from the results that the cocoamine ethoxylate does not show good performance in presence of anionics, particularly LAS. In combination with nonionics, amine ethoxylates perform well as dye transfer inhibitors.

Since the monoamine ethoxylates show good dye transfer inhibition proper-

TABLE 3 Effect of Surfactant Structure on Grass Stain
Removal (100°F, 150 ppm hard water, 2 g/L surfactant)

Surfactant	% Grass stain removed
LAS sodium salt	36.9
LAS isopropylamine salt	34.0
Nonylphenol ethoxylate (9 mole)	54.8
Alcohol ethoxylate (7 mole)	51.5
Tallowamine with 15 moles EO	70.0

Source: Ref. 8.

TABLE 4 Detergency Comparison of Amine and Alcohol Ethoxylates

	Percent soil removal	
	Cocoamine + 5 EO	C_{12-15} alcohol + 7 EO
Dust/sebum on cotton		
50°F	48.8	38.9
100°F	73.9	77.0
Dust/sebum on cotton/polyester		
50°F	59.0	49.9
100°F	62.0	62.3
Dust/sebum on polyester		
50°F	41.1	25.1
100°F	45.0	34.4
Olive oil on cotton		
50°F	70.0	71.5
100°F	79.7	82.1
Olive oil on cotton/polyester		
50°F	53.6	72.3
100°F	64.1	71.2
Olive oil on polyester		
50°F	19.2	1.3
100°F	49.2	4.7
Grass on cotton		
50°F	55.7	39.9
100°F	39.9	30.7

ties, the polyamine ethoxylates were expected to do better. Study of di-, tri-, tetra-, and pentaamine ethoxylates shows that more nitrogen in the molecule makes it a better dye transfer inhibition agent. The dye transfer inhibition data for polyamine ethoxylates are listed in Tables 8 and 9. It is obvious from the data that triamine ethoxylates are good surfactants for use as dye transfer inhibition

TABLE 5 Detergency of Amine and Alcohol Ethoxylates Blended with Anionics[a]

	Percent soil removed	
	Cocoamine + 5 EO	C_{12-15} alcohol + 7 EO
Dust/sebum on cotton		
LAS	81.9	71.3
SLES	85.0	84.1
SLS	85.1	83.9
Dust/sebum on cotton/polyester		
LAS	63.6	54.3
SLES	79.4	58.2
SLS	66.4	61.6
Dust/sebum on polyester		
LAS	46.4	29.8
SLES	37.4	30.7
SLS	45.0	33.9

[a]150 ppm hard water at 100°F, with 1.5 g anionic surfactant and 0.5 g nonionic surfactant per liter of wash water.
Source: Ref. 5.

agents. More than three nitrogen molecules do not add any performance benefits in the dye transfer inhibition area. For triamine ethoxylates it can be seen that the 10 mole EO adduct has the best performance. The triamine ethoxylate with 10 to 12 moles EO is a surfactant of choice for dye transfer inhibition performance in the laundry applications. Triamine ethoxylates are multifunctional products with detergency, dye transfer inhibition, antiredeposition, corrosion inhibition, and coupling properties. A comparison of the dye transfer inhibition properties of PVP and triamine ethoxylate is given in Table 10. Triamine ethoxylates have some cost/performance benefits over PVP, particularly because the triamine ethoxylates are multifunctional products with detergent as well as dye transfer inhibition properties. Owing to their detergent properties the formulator can take out an equal amount of nonionic surfactant from the formula when using triamine ethoxylates. The detergent properties of triamine ethoxylates are listed in Table 11. The data compares the detergent properties of AE and triamine ethoxylates. SLES and PVP are also included in the study. The results show that triamine ethoxylates perform as well as AE on the various soils included in the tests.

3. Formulations

Owing to their excellent detergent properties, ethoxylates amines can be used to formulate a variety of cleaning products both for household as well as industrial

TABLE 6   Correlation of Surface Properties with Cleaning

Composition (surfactant %)	Foam Height[a] (mL)	Wetting[b] (sec)	CMC (PPM)	Dynamic surf. ten.[c] (dynes/cm)	Detergency (average)
AE 100	36	29	11.3	56.4	40.2
AE 66.6 + AmE 33.3	39	51	24.6	56.2	40
AE 66.6 + AOS 33.3	42	37	12	61.7	38.1
AE 66.6 + LAS 33.3	40	27	13.5	61.3	38
AE 66.6 + SLES 33.3	40	39	11.9	60.4	41.3
AmE 100	36	300	12.5	52.6	39.2
AmE 66.6 + AE 33.3	36	243	9.2	54.7	42.1
AmE 66.6 + AOS 33.3	34	76	11.1	62.2	41.7
AmE 66.6 + LAS 33.3	33	64	7.3	63.6	37.2
AmE 66.6 + SLES 33.3	36	82	12.5	60.6	38.4
AOS 100	62	42	495	63.1	37.4
AOS 66.6 + AE 33.3	47	43	13.3	63.5	40
AOS 66.6 + AmE 33.3	34	41	10.6	65.7	39.1
AOS 66.6 + LAS 33.3	55	31	725	63.5	36.9
AOS 66.6 + SLES 33.3	60	39	415	61.1	41
LAS 100	58	16	854	63	23.5
LAS 66.6 + AE 33.3	55	22	14.3	63	34.6
LAS 66.6 + AmE 33.3	42	29	8.7	66.3	32
LAS 66.6 + AOS 33.3	62	20	572	62.9	32
LAS 66.6 + SLES 33.3	68	22	385	61	36.5
SLES 100	60	46	137.5	60.2	37.2
SLES 66.6 + AE 33.3	45	31	17.4	60.4	39
SLES 66.6 + AmE 33.3	36	47	13.9	62.6	38.4
SLES 66.6 + AOS 33.3	51	42	236	60.4	41
SLES 66.6 + LAS 33.3	64	28	231	60.3	38.8

[a]Foam height: Cylinder shake method. 0.01% active surfactant, 20 mL shaken in 100 mL graduated cylinder for 5 sec.
[b]Wetting: Drave's Wetting Test Using 0.05% active surfactant solution in DI water.
[c]Dynamic surface tension: 0.05% active surfactant solution in DI water was used in Sensadyne machine at 5 bubbles/sec.

and institutional markets. Some sample starting formulations are listed in Tables 12 through 18.

C. Conclusions

Coco and tallow amines with 5 to 20 moles (x + y) EO behave as nonionic detergents. A typical detergent nonionic AE has an HLB of 12.2 to 13.3 for the 7- and

TABLE 7 Dye Transfer Inhibition Dosage Curve for Amine Ethoxylate

Surfactant	Dosage	Avg. delta E
blank		31.9
AmE	0.025 g/L	30.6
	0.05 g/L	27.3
	0.1 g/L	23.7
	0.2 g/L	16.2
	0.4 g/L	11.4
	0.8 g/L	8.2
LAS	0.6 g/L	32.3
LAS + AmE	0.54 + 0.06 g/L	31.9
LAS + AmE	0.30 + 0.30 g/L	28.7
SLES (60%)	1.0 g/L	26.9
SLES + AmE	0.9 + 0.06 g/L	26.7
SLES + AmE	0.5 + 0.30 g/L	27.2
AE	0.6 g/L	19.3
AE + AmE	0.54 + 0.06 g/L	15.0
AE + AmE	0.30 + 0.30 g/L	11.0

TABLE 8 Dye Transfer Inhibition Properties of Di- and Triamine Ethoxylates at 0.1 g/L

Surfactant	Avg. delta E
Coco diamine with 3 EO	26.0
Coco diamine with 5 EO	21.8
Coco diamine with 10 EO	21.1
Coco diamine with 15 EO	22.0
Tallow triamine with 4 EO	17.9
Tallow triamine with 10 EO	11.3
Tallow triamine with 15 EO	13.2
Tallow triamine with 20 EO	13.4
Tallow triamine with 10 EO 0.3 g + LAS 0.3 g	26.9
Tallow triamine with 10 EO 0.48 g + LAS 0.12 g	11.4

9-mole ethoxylates, respectively. Ten moles of EO on tallow amine has a calculated HLB of 12.8, while 5 moles on cocoamine gives an HLB of 11.2.

From certain considerations, ethoxylated amines should be better surfactants than AE. A general rule is that a blend of two similar surfactants will perform better than a single one, and a blend of three is better than two, etc. This theory

TABLE 9 Dye Transfer Inhibition Properties of Tetra-
and Pentamine Ethoxylates with 10 Mole EO

Tallow tetramine with 10 EO	0.01 g/L	25.7
	0.025 g/L	22.3
	0.05 g/L	17.8
	0.10 g/L	13.5
Tallow pentamine with 10 EO	0.01 g/L	28.2
	0.025 g/L	17.7
	0.05 g/L	10.0
	0.1 g/L	14.0
PVP	0.01 g/L	24.9
	0.025 g/L	22.8
	0.05 g/L	19.8
	0.1 g/L	16.9

TABLE 10 Dye Transfer Inhibition of Tallowtriamine 12 Mole
Ethoxylate (TT-12EO) vs. PVP

Surfactant/polymer	Avg. delta E
PVP 0.01 g	26.0
PVP 0.02 g	23.7
PVP 0.05 g	20.2
PVP 0.1 g	17.3
TT-12EO 0.01 g	25.2
TT-12EO 0.02 g	24.6
TT-12EO 0.05 g	18.6
TT-12EO 0.1 g	16.3
LAS 0.5 g active	30.7
LAS 0.5 g active + PVP 0.02 g	29.8
LAS 0.5 g active + TT-12EO (pH 5.5) 0.02 g	30.8
LAS 0.5 g active + TT-12EO (pH 5.5) 0.05 g	30.4
SLES 0.5 g active	29.1
SLES 0.5 g active + PVP 0.02 g	22.3
SLES 0.5 g active + TT-12EO (pH 5.5) 0.02 g	28.6
SLES 0.5 g active + TT-12EO (pH 5.5) 0.05 g	25.2
AE 0.5 g active	19.0
AE 0.5 g + PVP 0.02 g	14.1
AE 0.5 g + TT-12EO (pH 5.5) 0.02 g	12.4
AE 0.5 g + TT-12EO (pH 5.5) 0.05 g	8.3
AE 0.4 g + SLES 0.1 g	27.4
AE 0.4 g + SLES 0.1 g + PVP 0.02 g	21.4
AE 0.4 g + SLES 0.1 g + TT-12EO (pH 5.5) 0.02 g	22.2

TABLE 11 Detergency Performance of Tallow Triamine 12 mole Ethoxylate (TT-12EO) on Dust Sebum, Olive Oil, and Grass Stain on Cotton and Cotton Polyester

Surfactant	Overall avg.
TT-12EO pH 5.5 0.5 g	56.3
AE 0.5 g	47.1
AE 0.45 g + TT-12EO 0.05 g	47.9
AE 0.4 g + SLES 0.1 g	44.1
AE 0.4 g + SLES 0.1 g + PVP 0.02 g	44.4
AE 0.38 g + SLES 0.1 g + TT-12EO 0.02 g	44.3
AE 0.35 g + SLES 0.1 g + TT-12EO 0.05 g	49.9

TABLE 12 Laundry Prespotter

Water	80.0%
LAS 100% active	12.0%
AmE	3.0%
AE	5.0%
Preservative, perfume, and dye	QS

TABLE 13 Laundry Prespotter

Water	80.0%
AmE	14.0%
AE	6.0%
Preservative, perfume, and dye	QS

TABLE 14 Heavy-Duty Liquid Laundry Detergent

Water	60.0%
Propylene Glycol	4.0%
SLES 60%	20.0%
AmE	10.0%
AE	6.0%
Preservative, perfume, and dye	QS

TABLE 15 Fine-Fabric Wash

Water	65.0%
SLES 60%	20.0%
AmE	10.0%
Cocoamidopropyl betaine	5.0%
Preservative, perfume, and dye	QS

TABLE 16 Carpet Cleaner

Water	81.0%
Sodium citrate	5.0%
Sodium lauryl sulfate 28%	6.0%
AmE	3.0%
Sodium xylene sulfonate	5.0%
Preservative, perfume, and dye	QS

TABLE 17 Tub and Tile Cleaner

Water	87.0%
Citric acid	8.0%
AmE	5.0%
Preservative, perfume, and dye	QS

TABLE 18 Toilet Bowl Cleaner

Water	92.0%
EDTA	3.0%
AmE	5.0%

appears to hold until the components added are too far away from the target HLB. Thus the development of the narrow-range AE which perform better than the original wide-ranged products. Statistically, the ethoxylated amines have many more similar components or isomers than the AE because two chains are growing instead of one during ethoxylation.

Trace levels of unreacted alcohol cause problems in traditional nonionics while all ethoxylated amines have a minimum of two EOs prior to building the chain. Again, an advantage in theory. By also acting as dye transfer inhibitors, amine ethoxylates have a definite advantage over AE. The weak point of ethoxylated amines and most specialty surfactants is cost. Depending on chain length and amount of EO, commercial ethoxylated amines cost between 20% and 100% more than AE.

III. ETHOXYLATED AMIDES

A. Structure and Method of Preparation

The two most common types of alkanolamides, diethanolamides and monoethanolamides, are made by reacting a fatty acid or ester with diethanolamine (DEA) or monoethanolamine (MEA), respectively (Fig. 6) [11]. For the diethanolamides, $x + y = 2$ generally; for the monoethanolamides, $n = 5$ or more. The EO chain can be extended on the ethanolamides in the same way as the ethoxylated amines by using a basic catalyst and adding the desired amount of EO.

B. Detergency and Other Useful Properties

Alkanolamides made by condensing 1 mole of fatty acid or ester with 1 mole of DEA are called 1:1 amides. If the ester used is a triglyceride; the resultant glycerine is left in the final product. The glycerine can be beneficial if the amide is used in a personal-care product, since the glycerine is a humectant and also has a good feel on the skin or hair. These so called 1:1 amides are not very soluble in water. The length of alkyl chain and the degree of unsaturation are important factors in determining the solubility of the amide. They are used in shampoos, dishwashing and laundry liquids, car wash, and other detergents as viscosity builders, foam boosters, and foam stabilizers.

Alkanolamides made by condensing one mole of fatty acid or ester with 2 moles of DEA are called 2:1 amides. They are generally free-flowing liquids as compared to 1:1 amides, which can be solids. Owing to the excess DEA, the 2:1 amides have better water solubility and detergency but less viscosity building properties than their 1:1 counterparts. They are used as detergents, wetting agents, and viscosity builders in a variety of industrial and institutional applications. Owing to fears of nitrosomines, the 2:1 amides are less common in recent years.

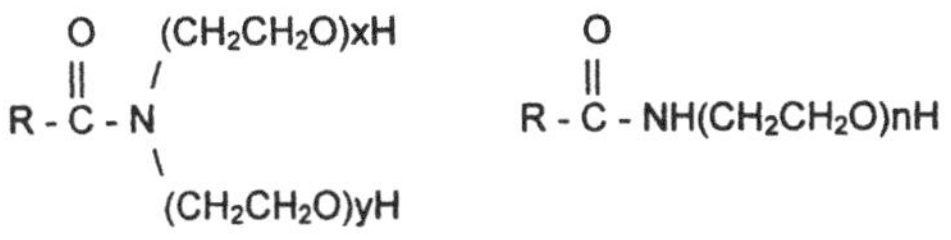

```
     O   (CH₂CH₂O)xH          O
     ‖  /                      ‖
  R - C - N              R - C - NH(CH₂CH₂O)nH
         \
          (CH₂CH₂O)yH
```

FIG. 6 Ethoxylated amides.

The monoethanol amides can be ethoxylated in order to increase their solubility in water. The ethoxylation raises their HLB and makes them better detergents and cleaners. The detergency performance of alkoxylated amides is shown in Table 19. The results listed in Table 19 point to the following conclusions:

1. As expected, it takes different amounts of ethylene oxide to optimize the detergency for cocamide MEA as compared to the tallowamide MEA. Tallow, on an average being longer chain than coconut requires more EO to have the solubility and HLB suitable for a detergent. Cocamide MEA, with 5 moles of EO, has enough hydrophilic character and solubility to be an effective surfactant. Tallowamide MEA, on the other hand, is not soluble at 5-mole EO adduct and has very poor detergency.

2. As the amount of EO is increased on cocoamide MEA, from 5 moles through 20 moles detergency improves on cotton with dust/sebum soil. It increases as the moles of EO are increased from 5 to 10 to 15 moles. Further increase from 15 to 20 moles of EO decreases the detergency (% Soil Removed). Cotton is a hydrophilic substrate and thus more soil is removed from cotton substrate as the surfactant become increasingly hydrophilic until a limit is reached. At 20 moles EO the surfactant becomes too soluble and its surface properties get negatively affected.

3. A similar phenomenon can be seen in case of tallowamide ethoxylates. The detergency is almost doubled as the moles of EO is increased from 5 to 10. Detergency further increases, although at much lower rate as the moles of EO are increased from 10 to 15. It finally drops off a little from the 15-mole to the 20-mole adduct.

4. Polyester is a hydrophobic substrate and that is why the relatively more hydrophobic surfactants in a series, clean polyester better. In cocamide MEA series, from 5 through 20 moles EO adducts the percentage soil removal for polyester soiled with dust/sebum decreases from 29.7 for 5-mole adduct to 23.3 for 10-mole adduct to 16.6 for 15 moles, and finally it reaches the lowest number of 14.3 for the 20-mole EO adduct. Similar behavior can be observed by looking into the percentage soil removal numbers for tallowamide MEA, where the numbers go down from 10 through 20 moles of EO.

5. Oily soils are removed by emulsification, solubilization, or rollover mechanisms. Surface tension, interfacial tension, and wetting properties of a surfactant play important roles in removing oily soil. Again a well-balanced HLB in a particular series of surfactants with similar structure will play an important part in determining the detergency performance of the surfactant on the oily soil. The data show that in the cocoamide MEA series the 5-mole adduct has the best detergency for oily soils, while for tallow, the 10-mole ethoxylate is the best performer, followed by the 15-mole and the 20-mole ethoxylate.

6. Addition of propylene oxide to alcohol ethoxylates has shown to lower the foam and improve the liquidity, and may also improve the detergency. Addition of

TABLE 19 Detergency of Ethoxylated Amides on Particulate Soils (dust sebum, standard soil, clay), Oily Soils, and Stains (chocolate milk, coffee, grape juice, grass, ketchup, tea) Using 1g/L at 100°F

	Soils	Avg. SRI	Avg. %SR
Cocamide MEA+5 EO	Particulate	78.3	37.3
Cocamide MEA+10 EO		77.6	36.0
Cocamide MEA+15 EO		77.1	34.4
Cocamide MEA+20 EO		76.6	32.9
Cocamide MEA+2 PO+5 EO		78.5	37.5
Cocamide MEA+2 PO+10 EO		77.7	36.0
Cocamide MEA+2 PO+15 EO		77.7	36.3
Cocamide MEA+2 PO+20 EO		77.0	34.1
Tallowamide MEA+5 EO		73.0	18.3
Tallowamide MEA+10 EO		77.6	35.2
Tallowamide MEA+15 EO		78.4	38.6
Tallowamide MEA+20 EO		77.3	35.4
Cocamide MEA+5 EO	Oily	93.4	73.8
Cocamide MEA+10 EO		84.5	38.0
Cocamide MEA+15 EO		85.4	42.3
Cocamide MEA+20 EO		85.3	41.7
Cocamide MEA+2 PO+5 EO		90.6	63.2
Cocamide MEA+2 PO+10 EO		86.7	47.5
Cocamide MEA+2 PO+15 EO		86.0	44.5
Cocamide MEA+2 PO+20 EO		86.1	44.7
Tallowamide MEA+5 EO		84.1	35.3
Tallowamide MEA+10 EO		87.9	53.1
Tallowamide MEA+15 EO		86.5	47.3
Tallowamide MEA+20 EO		85.8	42.9
Cocamide MEA+5 EO	Stains	86.8	51.6
Cocamide MEA+10 EO		88.7	62.4
Cocamide MEA+15 EO		86.2	52.7
Cocamide MEA+20 EO		85.6	48.8
Cocamide MEA+2 PO+5 EO		84.0	34.2
Cocamide MEA+2 PO+10 EO		88.4	56.9
Cocamide MEA+2 PO+15 EO		88.2	55.8
Cocamide MEA+2 PO+20 EO		88.0	55.9
Tallowamide MEA+5 EO		86.5	53.8
Tallowamide MEA+10 EO		87.7	56.7
Tallowamide MEA+15 EO		87.6	56.7
Tallowamide MEA+20 EO		87.7	57.1

TABLE 20 Average Percent Soil Removal for a Variety of Surfactants and Blends with Tallowamide MEA + 17.5 moles EO (Amd-EO)

Fabric	Stain	25–7 0.5 g	SLES 0.5 g	Amd-EO 0.5 g	25–7 0.3 g + Amd-EO 0.2 g	SLES 0.3 g + Amd-EO 0.2 g
Cotton	D/S	52.7	55.8	73.9	73.3	54
	Std.	14.9	19.7	22.2	28.8	18.6
	Clay	68.2	66.2	73.3	69.9	67
Cot/pol	D/S	77.9	71.1	67.6	79.6	68.9
	Std.	20.3	18.6	20.2	32.2	12.2
	Clay	78.7	78.7	80.5	79.7	77.7
Cot/pol	Olive oil	70.1	67.7	72.2	70.2	66.8
	Margarine	79.0	75.5	77.8	76.7	75.5
Cot/pol	Grass	33.8	24.3	33.5	35	26
	Blood	48.2	68.6	61.6	50.7	69.8
	Coffee	88.4	86.2	65.4	83.4	86.4
	C. milk	52.2	69.8	68.7	63.1	65.4
	G. juice	70.1	62.2	71.1	70.5	61.8

2 moles of propylene oxide to the cocamide MEA ethoxylates does not have any big effect on the detergency except for polyester fabric where the inclusion of propylene oxide has shown to improve the percentage removal of dust/sebum soil.

The detergency performance of amide ethoxylates in laundry application is well in line with the other commonly used surfactants in liquid laundry formulations like SLES and AE (Table 20).

C. Formulations

The formulations listed in Tables 21 through 26 illustrate the practical applications of alkanolamides and their ethoxylates. Amide ethoxylates, particularly tallow, are very cost-effective nonionics; however, they can be dark in color. Amide ethoxylates, like most nitrogen-containing surfactants, do possess dye transfer inhibition properties.

TABLE 21 Heavy-Duty Floor Cleaner

Water	78.0%
Cocamide DEA 2:1	7.0%
TKPP	10.0%
Sodium xylene sulfonate	5.0%
Preservatives, dye, perfume	QS

TABLE 22   Laundry Prespotter Gel

Water	81.0%
AE	8.0%
Sodium lauryl ether sulfate	6.0%
Oleamide DEA	3.0%
Sodium chloride	2.0%
Preservatives, dye, perfume	QS

TABLE 23   Laundry Detergent

Water	70.0%
LAS sodium salt	20.0%
AE	6.0%
Tallowamide MEA ethoxylate	4.0%
Preservatives, dye, perfume	QS

TABLE 24   Fine-Fabric Wash

Water	80.0%
SLES	10.0%
Cocamidopropyl betaine	4.0%
Tallowamide MEA ethoxylate	6.0%
Preservatives, dye, perfume	QS

TABLE 25   Industrial Degreaser

Water	65.0%
LAS sodium salt	10.0%
AE	5.0%
Glycol ether	10.0%
Cocamide DEA 2:1	5.0%
TKPP	5.0%
Preservatives, dye, perfume	QS

TABLE 26   Heavy-Duty D'Limonene Cleaner

Phase A	
D'Limonene	35.0%
Cocamide DEA	4.0%
Nonionic surfactant	13.0%
Glycol ether	6.0%
Phase B	
Sodium C_{14-16} Alpha olefin sulfonate	10.0%
Sodium naphthalene sulfonate	8.0%
Water	24.0%
Preservatives, dye, perfume	QS

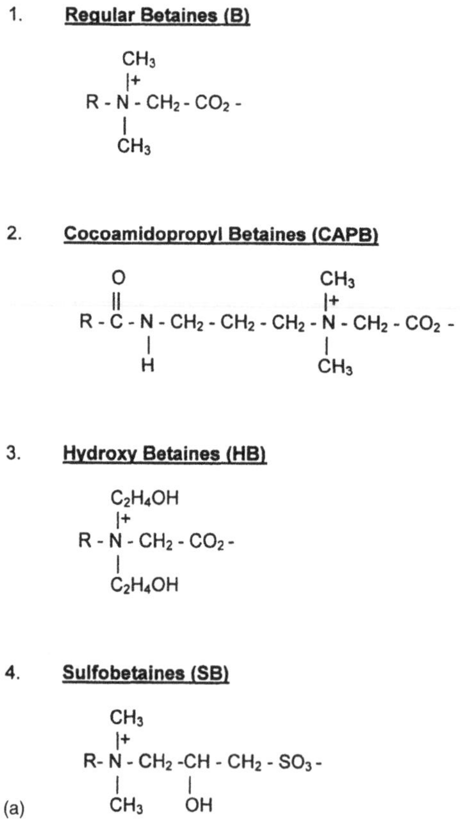

1. **Regular Betaines (B)**

$$CH_3$$
$$|+$$
$$R - N - CH_2 - CO_2 -$$
$$|$$
$$CH_3$$

2. **Cocoamidopropyl Betaines (CAPB)**

$$O \qquad\qquad CH_3$$
$$|| \qquad\qquad\quad |+$$
$$R - C - N - CH_2 - CH_2 - CH_2 - N - CH_2 - CO_2 -$$
$$| \qquad\qquad\qquad |$$
$$H \qquad\qquad\quad CH_3$$

3. **Hydroxy Betaines (HB)**

$$C_2H_4OH$$
$$|+$$
$$R - N - CH_2 - CO_2 -$$
$$|$$
$$C_2H_4OH$$

4. **Sulfobetaines (SB)**

$$CH_3$$
$$|+$$
$$R- N - CH_2 -CH - CH_2 - SO_3 -$$
$$| \qquad |$$
(a) $\qquad CH_3 \quad OH$

FIG. 7   Amphoterics and betaines.

IV. AMPHOTERICS AND BETAINES

A. Structure and Method of Preparation

Seven structural types of amphoterics and betaines are commonly known and used (Fig. 7). Structures B, CAPB, and HB are made by reacting sodium chloroacetate (SCA) with an appropriate substituted tertiary amine usually in water (Fig. 8). A mole of salt is produced, some of which precipitates out and is removed by filtration, with the rest remaining in solution. For B, a commercially available alkyldimethylamine is used, while for structure HB an ethoxylated alkylamine is the building block. For CAPB, an amidoamine is first prepared from dimethylaminopropylamine (DMAPA) and either a fatty acid or ester.

Sulfobetaines SB and CAS are similar to betaine B and CAPB except that sodium 3-chloro-2-hydroxy-propanesulfonate is the reactant in place of sodium chloroacetate. This material is prepared just prior to use from sodium bisulfite and epichlorohydrin (Fig. 9). All these betaine and sulfonbetaine reactions produce salt as a byproduct, which generally is just left in the product. Commercially, these products are sold as 30% to 40% aqueous solutions.

5. **Cocoamidopropyl Sulfobetaines (CAS)**

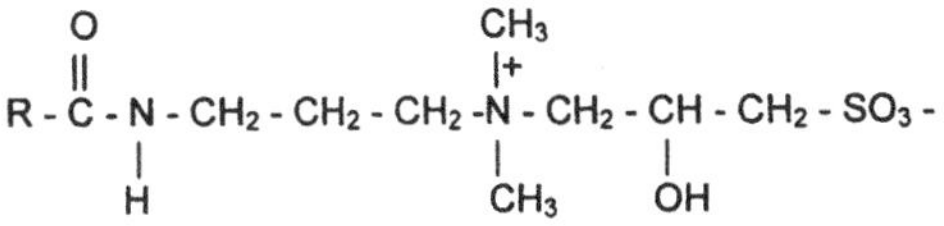

$$
\begin{array}{ccc}
O & & CH_3 \\
\parallel & & \mid^+ \\
R-C-N-CH_2-CH_2-CH_2-N-CH_2-CH-CH_2-SO_3- \\
\mid & & \mid \quad\quad \mid \\
H & & CH_3 \quad OH
\end{array}
$$

6. **Imidazoline Derived Amphoterics (AM)**

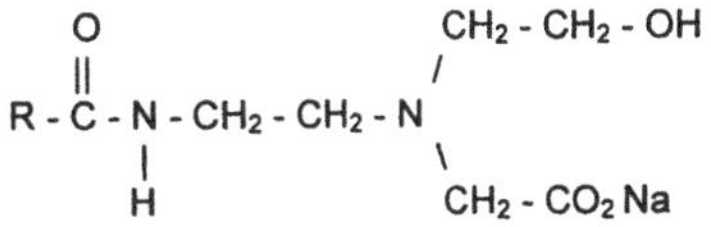

$$
\begin{array}{cc}
O & CH_2-CH_2-OH \\
\parallel & \diagup \\
R-C-N-CH_2-CH_2-N \\
\mid & \diagdown \\
H & CH_2-CO_2\,Na
\end{array}
$$

7. **Imidazoline Derived Amphoterics - Salt Free (AM-SF)**

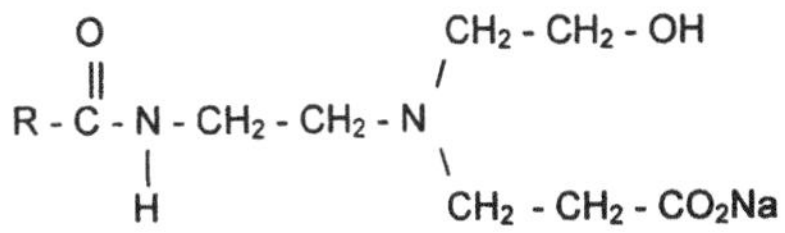

$$
\begin{array}{cc}
O & CH_2-CH_2-OH \\
\parallel & \diagup \\
R-C-N-CH_2-CH_2-N \\
\mid & \diagdown \\
H & CH_2-CH_2-CO_2Na
\end{array}
$$

(b)

FIG. 7 Continued

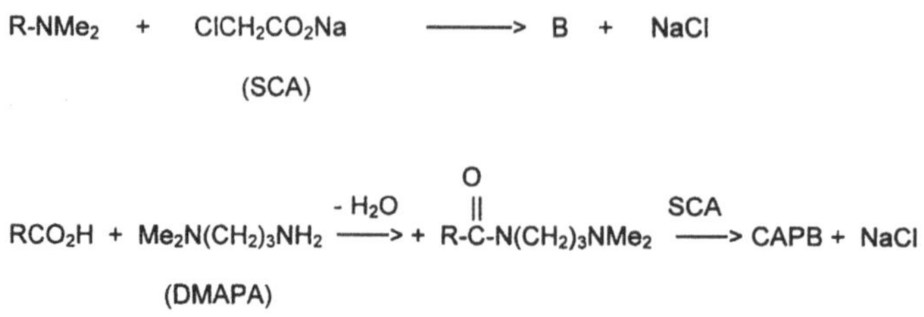

R-NMe$_2$ + ClCH$_2$CO$_2$Na $\longrightarrow$ B + NaCl

        (SCA)

$$RCO_2H \ + \ Me_2N(CH_2)_3NH_2 \ \xrightarrow[]{-H_2O} \ + \ R\overset{\overset{\displaystyle O}{\|}}{C}\text{-}N(CH_2)_3NMe_2 \ \xrightarrow{SCA} \ CAPB \ + \ NaCl$$

        (DMAPA)

FIG. 8 Synthesis of betaines.

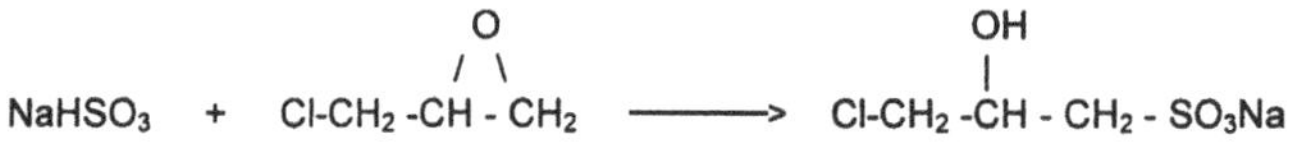

$$NaHSO_3 \ + \ Cl\text{-}CH_2\overset{\overset{\displaystyle O}{/ \ \backslash}}{\text{-}CH\text{-}}CH_2 \ \longrightarrow \ Cl\text{-}CH_2\overset{\overset{\displaystyle OH}{|}}{\text{-}CH\text{-}}CH_2\text{-}SO_3Na$$

FIG. 9 Reaction of sodium bisulfite and epichlorohydrin.

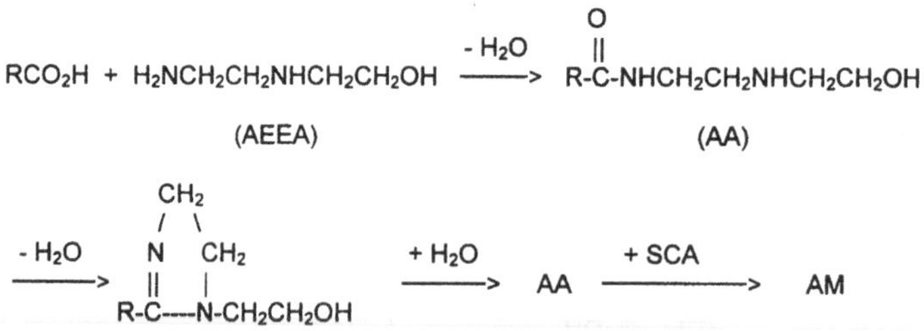

$$RCO_2H \ + \ H_2NCH_2CH_2NHCH_2CH_2OH \ \xrightarrow[]{-H_2O} \ R\overset{\overset{\displaystyle O}{\|}}{C}\text{-}NHCH_2CH_2NHCH_2CH_2OH$$

        (AEEA)   (AA)

$$\xrightarrow[]{-H_2O} \ \begin{matrix} CH_2 \\ / \ \backslash \\ N \ \ \ CH_2 \\ \| \ \ \ | \\ R\text{-}C\text{----}N\text{-}CH_2CH_2OH \end{matrix} \ \xrightarrow[]{+H_2O} \ AA \ \xrightarrow[]{+SCA} \ AM$$

FIG. 10 Synthesis of imidazoline-based amphoteric.

The first step in synthesizing amphoterics AM and AM-SF is to form an amidoamine followed by an imidazoline from aminoethylethanolamine (AEEA) and either a fatty acid or ester (Fig. 10). However, during reaction with aqueous sodium chloroacetate the imidazoline hydrolyzes back to the amidoamine (AA) then reacts with the sodium chloroacetate [12]. The salt-free amphoteric (AM-SF) follows a similar path through the imidazoline and its hydrolysis except that the AA is reacted with methyl acrylate and then caustic to hydrolyze the ester.

True amphoteric surfactants (AM, AM-SF) contain both the positive and negative groups in the same molecule, which leads to their unique and useful properties. In acidic pH an amphoteric is predominantly cationic, whereas in alkaline pH it is anionic. The ionic nature of amphoteric surfactants therefore depends on the pH. At an intermediate pH the strength of both positive and negative charge will get equal and the molecule will have positive as well as negative charge. At this point it is a zwitterion. The pH at which both charges are equal in strength is called the isoelectric point.

The isoelectric point is not a sharp point but depends on the nature of the anionic and cationic groups. At the isoelectric point amphoterics generally have their minimum solubility.

Betaines contain a quaternary nitrogen atom in their molecule (essentially a permanent positive charge), which makes them different from other amphoterics. Owing to the quaternary nitrogen a betaine cannot exist in anionic form, even at alkaline pH. Betaines only can exist in cationic or zwitterionic forms.

Sulfobetaines contain a sulfonate group instead of the carboxyl group. The sulfonate group makes a sulfobetaine more hydrophilic and more tolerant to the high pH than the carboxyl group betaines.

Any of the structures derived from sodium chloroacetate can be called glycinates, since they are derivatives of the aminoacid glycine. They can also be called acetates although that name is more common for the amphoteric AM. There are two forms of the AM—the one shown in Figure 7 where 1 mole of SCA is used, and the diacetate where a second mole of SCA converts the AM to a betaine. The AM-SF can be called a propionate as it is a derivative of propionic acid.

B. Detergency and Other Useful Properties

1. Applications of Amphoteric and Betaine Surfactants

Amphoteric/betaine surfactants have been used in a wide variety of products including household, I&I, and personal-care detergents [13,14]. Some of their advantages include mildness, stability in acidic and alkaline formulations, hydrotroping and coupling properties, synergism with other surfactants, foam stabilization, and viscosity modification.

2. Amphoterics in Laundry Products

Compared to the commonly used anionic and nonionic surfactants, amphoterics/betaines are generally mild, with lower skin and eye irritation. Amphoterics/betaines can therefore be used in laundry products where mildness is an important criteria, like laundry detergents for baby clothing and for sensitive skin people. Fine-fabric wash formulas can also benefit from addition of amphoteric/betaine surfactants. Laundry bar formulations can incorporate amphoteric/betaine surfactants in order to improve the mildness on hands.

Amphoteric/betaine surfactants can be used in I&I detergents because of extreme pH stability. In institutional laundry detergents, where relatively high amounts of caustic are used, amphoteric/betaine surfactants are well suited because of their alkali stability. Softergent (detergent and fabric softener in one formula) formulations require surfactants that are compatible with quats.

Amphoterics/betaines are ideal here because of their excellent quat compatibility. In fabric softener formulations, amphoterics/betaines can be used as dispersing agents for the quat.

The detergency of CAPB on oily soils was compared to that of a commonly used AE. The results are listed in Table 27 showing good soil removal for the betaine.

Betaines can also be used in laundry detergents as dye transfer inhibitors for acidic and direct dyes. The cationic nature of the nitrogen in a betaine like CAPB allows it to adsorb on the negatively charged dye molecules and form a complex. The betaine acts as a fixative agent for the dye and resists the migration or the transfer of the dye. The dye transfer inhibition performance of CAPB together with other surfactants was evaluated. The lower delta E values mean less dye transfer, and thus the lower values are desirable and are listed in Table 28. A quick look at the results listed in Table 28 shows that the nitrogen-containing surfactants act as strong dye transfer inhibitors. Anionic surfactants, on the other hand, act as dye transfer promoters. The anionics can

TABLE 27 Detergency of CAPB and an AE on Eight Oily Soils

	Ave. % soil removed			
	CAPB		AE	
Fabric	50°F	70°F	50°F	70°F
Cotton	55.1	83.8	51.5	82.2
Cot/pol	54.0	73.0	50.6	77.0
Polyester	17.1	54.1	8.1	48.4

TABLE 28 Dye Transfer Inhibition of Detergent Formulas[a]

Formulas	A	B	C	D	E
Water	55	55	55	55	55
AmE	15	15	15	15	15
AE	20	20	20	20	20
LAS	10				
SLES		10			
Sodium C_{14-16} alpha olefin sulfonate			10		
Carboxylated C_{13} alcohol with 6 EO				10	
CAPB					10
Resulting ave. delta E	21.7	21.1	23.4	8.4	8.1

[a]All quantities are weight percents on 100% active basis.

solubilize the acid and direct dyes in their micelles and promote migration of the dye.

CAPB has excellent dye transfer inhibition properties. This surfactant can be used in laundry detergent formulas, where mildness and dye transfer inhibition are important performance criteria. The formulated blends A through E can act as starting-point formulation guidelines for such products. Fine-fabric washes and laundry prespotters can also be formulated using these blend compositions as starting point. Examples of the finished laundry product formulations are in Tables 29 through 34.

Amphoterics/betaines can also be used in I&I laundry detergents where high alkalinity is common to the formulations. Sulfobetaines or Sultaines are particu-

TABLE 29   Mild Liquid Laundry Detergent

Water	to 100%
Propylene glycol	5.0%
Enzymes	QS
Sodium Citrate	5.0%
CAPB	10% actives
AmE	10% actives
SLES	5.0% actives
Optical brightners	QS
Dyes, preservatives	QS

TABLE 30   Fine-Fabric Wash

Water	to 100%
SLES	8.0% actives
CAPB	4.0% actives
Dimethyllauramine oxide	4.0% actives
Propylene glycol	5.0
Dyes, preservatives	QS

TABLE 31   Softergent

Water	to 100%
Oleyl imidazoline quat	8.0% actives
CAPB	10.0% actives
AmE	10.0% actives
Propylene glycol	5.0
Dyes, preservatives	QS

TABLE 32 Laundry Prespotter

Water	to 100%
CAPB	5.0% actives
AE	8.0% actives
Dyes, preservatives	QS

TABLE 33 I&I Liquid Laundry Detergent

Water	70.0%
Sodium metasilicate pentahydrate	5.0%
Tetrapotassium pyrophosphate	10.0%
Potassium hydroxide	5.0%
CAS	10.0% actives
Preservatives, dye	QS

TABLE 34 I&I Liquid Fabric Softener

Water	QS to 100%
Dihydrogenated tallow dimethyl ammonium chloride	8.0% acvtives
AM-SF	2.0% actives
Hexylene glycol	3.0%

larly suitable for high-alkaline formulations. Amphoterics can also act as hydrotropes and solubilize other surfactants used in the formulation.

3. Amphoterics in Hard-Surface Cleaning

(a) Hand Dishwash Detergents. Hand dishwash liquids are formulated to give a high flash foam together with good foam quality and foam stability. The grease-cutting properties are also very important for a dishwash detergent. Mildness to the skin is another important criteria that formulators look for in a hand dishwash formula. Surfactants with viscosity-building properties are also chosen for use in liquid dishwash detergents. Betaines can provide several performance advantages including mildness, foam stability, viscosity building, and detergency. Other amphoterics like AM and AM-SF can also be used in dishwash formulas for their mildness and detergent properties. Some starting-point formulations for hand dish detergents using amphoteric surfactants are listed in Tables 35 and 36.

(b) All-Purpose Cleaners/Degreasers. Amphoterics/betaines work well in high- and low-pH formulas as well as in products containing high amounts of

TABLE 35   Mild Hand Dishwash

Water	to 100%
SLES	20.0% actives
CAPB	6.0% actives
Dye, fragrance, preservatives	QS

TABLE 36   Mild Hand Dishwash

Water	to 100%
LAS	12.0%
SLES	8.0%
AM-SF	10.0% actives
Viscosity control agent (alcohol or SXS)	QS
Dye, fragrance, preservative	QS

TABLE 37   Ceramic Tile Soil

Lard	30%
Crisco vegetable shortening	10%
Wesson vegetable oil	10%
Brandy black clay	25%
Potting soil	25%

electrolyte. Many all-purpose cleaners or kitchen degreaser formulas are based on high-alkali contents. The ingredients include sodium or potassium hydroxide, phosphates, silicates, carbonates, etc. When nonionic surfactants like nonylphenol ethoxylates or alcohol ethoxylates are used, the formula requires a high percentage of hydrotropes like Sodium Xylene Sulfonate (SXS). Amphoterics/betaines can be used in this type of formulations since they act as hyrotropes as well. Their use may also give a cost/performance advantage to the formulator.

Experiments were run in order to demonstrate the performance of amphoteric surfactants in all purpose cleaners. The soil was prepared according to Table 37 and applied to ceramic tile. A base builder formula was prepared as listed in Table 38, and the experimental formulas were prepared according to the compositions in Tables 39 and 40.

The cleaner was applied by dropping 100-μL from a multitip pipette onto a soiled ceramic tile. A series of drops were placed at intervals of 2 sec. At the end of 20 sec, the tile was rinsed with tap water and a slow stream of water was directed onto the spot where the drop was. The following conclusions were made after the test was completed:

TABLE 38   Base Builder Formula for
Ceramic Tile

Water	80.0%
Tetra potassium pyrophosphate	12.5%
Sodium metasilicate pentahydrate	3.5%
Sodium hydroxide	4.0%

TABLE 39   Test Formulas for Cleaning Ceramic Tile

	Formula No.			
	1	2	3	4
Base builder formula	60.0%	60.0%	60.0%	60.0%
Nonyl phenol ethoxylate (9 mole)	6.0%			
Sodium xylene sulfonate	15.0%			
SB		6.0%		
AM-SF			6.0%	
AM				6.0%

TABLE 40   Test Formulas for Cleaning Ceramic Tiles with NPE-9

	Formula No.		
	5	6	7
Base builder formula	60.0%	60.0%	60.0%
Nonyl phenol ethoxylate (9 mole)	3.0%	3.0%	3.0%
Sodium xylene sulfonate			
SB	6.3%[a]		
AM-FS		8.0%[a]	
AM			11.0%[a]

[a]Minimum amount to produce a clear formula.

The base formula did not clean by itself.

The base formula with NPE and SXS did not clean well and it was unstable (separated in layers).

Surfactants SB and AM-SF both exhibited some cleaning. The cleaning was much better when the nonionic was used with the amphoteric.

Formulas were more stable when an amphoteric was used.

The above experiment demonstrated the performance advantages of amphoterics in high-alkali all-purpose cleaner/degreaser formulas. Examples of starting formulas for all purpose cleaners and degreasers are given in Tables 41 through 44.

(c) Soap Scum Removal.   Bathroom tub and tile cleaners are formulated to remove soap scum from the tub and tiles. Soap scum is a mixture of soap, hard water, shampoos, sebum, dirt, etc. Tub and tile cleansers are generally acidic or

TABLE 41   All-Purpose Cleaner

Water	90.0%
Sodium carbonate	2.0%
Sodium tripolyphosphate	3.0%
Sodium metasilicate pentahydrate	2.0%
AE	1.0%
SB	2.0%

TABLE 42   All-Purpose Cleaner

Water	80.0%
Tetrapotassium pyrophosphate	5.0%
AM-SF	3.0%
Propylene glycol butyl ether	6.0%
C8 paraffin sulfonate	3.5%
Sodium alkyl naphthalene sulfonate	2.5%

TABLE 43   All-Purpose Disinfectant Cleaner[a]

Water DI	92.0%
Dimethyl alkyl (C_{12-16}) benzyl ammonium chloride	5.0%
CAPB	3.0% actives

[a]This formula is intended as a starting-point suggested formula. No lab work has been done on the formula. Perform all necessary tests and comply with all regulations. EPA/FDA registration may be required if sold as an antibacterial product.

TABLE 44   All-Purpose Degreaser

Water	81.0%
Sodium metasilicate pentahydrate	5.0%
EDTA	2.0%
Propylene glycol butyl ether	4.0%
Sodium alkyl naphthalene sulfonate	5.0%
AM	3.0%

alkaline products, not neutral. Amphoterics are well suited in these formulations because of their wide range of pH compatibility.

Tests were conducted to evaluate the performance of "soap scum removers" with the ingredients listed in Tables 45 and 46. The soap scum was prepared by adding the hard water to the soap solution, graphite, isopropanol, and sebum, and spraying on Formica panels. For the test, two drops of the test formulation were applied (see Table 47) to at least four areas of the soiled Formica tile panel. After waiting for 30 sec, the panels were rinsed and evaluated using a scale from 0 (worst) to 4 (best) to rate the soil removal of each test solution. The results are given in Table 48.

TABLE 45   Synthetic Sebum

Ingredients	% by weight
Palmatic acid, reagent grade	10.0%
Stearic acid powder, triple pressed	5.0%
Coconut oil	15.0%
Paraffin wax	10.0%
Sperm wax	15.0%
Olive oil	20.0%
Squalene	5.0%
Cholesterol	5.0%
Oleic acid	10.0%
Linoleic acid	5.0%

Source: Ref. 15.

TABLE 46   Soap Scum Removal Test Materials

Equipment/reagents	
Hard water	1.0 L DI water
	10.0 g calcium acetate
	3.0 g magnesium nitrate
Soap solution	1.0 L DI water
	100.0 g Ivory soap chips
Isopropyl alcohol	
Grade #38 graphite powder	
Test formulations	
Matte finish white formica	
Spray tool	

TABLE 47 Test Formula for Soap Scum
Removal

Amphoteric/betaine surfactant	1.0% active
Sodium EDTA	0.25%
Ethylene glycol n-butyl ether	0.50%
Builder solution[a]	0.20%
Water	to 100%

[a]Builder solution: potassium hydroxide, 3%; sodium
carbonate, 4%; water, 97%.

TABLE 48 Soap Scum
Removal Results

SB	50% soil removed
CAS	75% soil removed
AM-SF	50% soil removed
C_8 AM	100% soil removed
CAPB	25% soil removed

0	no removal
1	25 % soil removed
2	50 % soil removed
3	75 % soil removed
4	100 % soil removed

Some starting point formulas for bathroom tub and tile cleaners are listed in Tables 49 through 52.

(d) Nonstreak Cleaning. The fundamental tendency of a surfactant to leave a residue on a surface is a very important parameter to consider during formulation of glass and multisurface cleaners. The residue left behind after cleaning might

TABLE 49 Bathroom Tile Cleaner

Water	90.0%
AM-SF	3.0%
Tetrasodium EDTA	4.0%
Ethylene glycol monobutyl ether	3.0%
Dye, preservative, perfume	QS

TABLE 50   Tub and Tile Cleaner

Water DI	to 100%
Xanthan gum	0.5%
CAS	2.5%
Phosphoric acid 75%	10.0%
Dye, preservative, perfume	QS

TABLE 51   Tub and Tile Cleaner

Water DI	to 100%
Phosphoric acid 75%	10.0%
Hydrochloric acid 37%	5.0%
HB	2.0%
Tallowamine ethoxylate (2 mole EO)	1.0%
Dye, preservative, perfume	QS

TABLE 52   Tub and Tile Cleaner

Water	to 100%
Tetra potassium pyrophosphate	3.0%
Potassium hydroxide	0.5%
CAS	1.5%
Propylene glycol monobutyl ether	3.0%
Sodium alkyl naphthalene sulfonate	3.0%
Dye, preservative, perfume	QS

be exhibited as a film or streaks on the freshly cleaned surface. This filming or streaking is very undesirable in consumer cleaners.

Surfactants were evaluated for their streaking and filming properties by applying 2 mL of test surfactant solutions to mirrored panels and wiping off with a paper towel and graded on a scale of 0 to 6 with 6 equaling to no streaks present:

Grade 0	A grade 0 is given to the products which yield exceedingly prominent streaking. The streaking characteristics are white-colored and very "frothy" looking. The streaking is easily seen and the surface is 100% affected.
Grade 1	Streaking is very prominent and uniform; 100% of the surface is affected. Streaking is beginning to take a white frothy appearance. Streaking is easily seen.
Grade 2	Streaking is uniform but less prominent. Entire mirror surface is affected and the streaking is easily seen.

Grade 3 Light uniform streaking is observed. More than 75% of the surface is affected. Streaking is not as evident. Proper viewing angles are required.

Grade 4 Streaking is spotty, with approximately 50% of the surface being affected. The streaking is not prominent.

Grade 5 Minimal streaking is observed. Streaking is very limited and localized and usually in a single area. The majority of the surface is clean and free from streaking.

Grade 6 No streaking is observed. Mirror surface is in the same condition as prior to application.

A number of surfactants were evaluated for their streaking behavior, using the above test procedure. A 0.25% active surfactant solution in DI water was used for the test. The results are as listed in Table 53. Some suggested starting point formulas for glass and other light duty all purpose clears are given in Tables 54 and 55.

TABLE 53 Glass Streaking Results

Surfactant	Streaking grade
Sodium lauryl ether sulfate	3.5
Ammonium lauryl sulfate	2.8
Nonylphenol (9 mole) ethoxylate	1.5
Cocoamine (10 mole) ethoxylate	3.0
Linear alkylbenzene sulfonate	3.3
Sodium alpha olefin sulfonate	2.8
Lauryl dimethyl amine oxide	3.0
Alcohol ethoxy (6 mole) carboxylate	2.8
Sodium alkyl naphthalene sulfonate	2.5
AM-SF	4.5
CAPB	3.3
C_8 AM	3.5
SB	3.5
CAS	5.3

TABLE 54 Solvent-Free Glass and Multipurpose Cleaner

DI water	to 100%
CAS	4.0%
Ammonia 29%	2.0%
Dye, preservative, perfume	QS

TABLE 55 Glass Cleaner

Water	to 100%
Isopropyl alcohol	5.0%
Ethylene glycol monobutyl ether	3.0%
Ammonia 29%	2.0%
AM-SF	1.0%

(e) Hydrotroping and Coupling Properties of Amphoteric/Betaine Surfactants. The recent trend of high active laundry, hard-surface, and other household cleaners has increased the use of hydrotropes, solubilizers, and coupling agents in the formulations. Hydrotropes or couplers are used to solubilize the active organic components in water. The most popular hydrotrope has traditionally been SXS. SXS is an adequate coupler but it does not generally contribute any cleaning performance to the formulation. Amphoterics/betaines, on the other hand, are multifunctional when used in cleaning applications. They not only contribute to the detergency but also provide coupling properties to the formulation.

The coupling property of various hydrotropes were evaluated using a special test procedure. In the procedure 100 mL of a mixture which is a two-layer blend of builders and organic surfactants (see Table 56) is stirred in a beaker while slowly adding test surfactants drop by drop to the beaker until the system gets clear and stays clear for at least 3 min. The effectiveness of the test surfactants were calculated as in Figure 11 and rated using Table 57.

Using this test procedure and rating system, a number of amphoteric/betaine surfactants were rated for their coupling performance. The results are listed in Table 58 and show that amphoteric surfactants can act as excellent hydrotropes. They are much more effective than SXS, but they more costly.

(f) Acid Thickening. Bathroom cleaners like toilet bowl cleaners and tub and tile cleaners may contain a high percentage of mineral or organic acids. To achieve the vertical cling of these products on a ceramic surface, a high viscosity is required. Polymers like carboxymethyl cellulose and others do not work

TABLE 56 Coupling Test Mixture

Component	% W/W
DI water	64.0%
Sodium metasilicate pentahydrate	5.0%
Tetrapotassium pyrophosphate	5.0%
Nonylphenol ethoxylate	2.0%
Potassium hydroxide 30%	24.0%

$$\frac{\text{Activity of the coupler (As is)} \times \text{Grams of coupler used}}{100} = S$$

$$\frac{S}{100 + \text{grams of coupler used}} \times 100 = C$$

$$C \times 3 = Z$$

FIG. 11 Formulas to determine activity of a coupler.

TABLE 57 Coupling Test
Rating Values

If "Z" is:	Rating is:
< 1	10 (best)
1–1.9	9
2–2.9	8
3–3.9	7
4–4.9	6
5–5.9	5
6–6.9	4
7–7.9	3
8–8.9	2
9–9.9	1
10 +	0 (worst)

TABLE 58 Coupling Test Ratings

Surfactant	Coupling rating
Sodium xylene sulfonate (SXS)	3
Paraffin (C_8) sulfonate	5
Paraffin (C_{10}) sulfonate	6
CAPB	6
C_8 AM	7
SB	8
AM-SF	8
AM	8
CAS	8

well for thickening the acids. One of the products that does work very well is dihydroxy tallow glycinate (HB). It is effective in thickening both organic and inorganic acids. Thickened acid formulations can be found in many different applications. Some of the more common uses include metal cleaning, transportation cleaners, tub and tile cleaners, aluminum brighteners, and cement etching solutions.

A stable viscosity at elevated temperature is one of the primary advantages of a thickened acid solution using dihydroxy tallow glycinate. A sample formula as listed in Table 59 was prepared and subjected to higher temperatures in order to evaluate the heat stability of the formula. The formula was subjected to 52°C and periodically checked for viscosity. The results are listed in Table 60.

HB will thicken hydrochloric acid by itself. Other acids such as phosphoric acid would require the addition of an electrolyte together with the glycinate in order to increase the viscosity. Different viscosity responses can be obtained by changing the salt. In order to evaluate the viscosity response with various salts, a test following formula was prepared (see Table 61). Several different salts were added to the above formula in order to increase the viscosity. The results of viscosity response using various salts are listed in Table 62.

TABLE 59   Thickened Acid Cleaner

Water	75.4%
HB	6.0%
Hydrochloric acid 37%	27.0%

TABLE 60   Heat Stability Test Results of Acid-Thickened Formula

Days at 52°C	Initial	3	5	11	21
Viscosity (cps)	2080	2040	1860	2040	1800

TABLE 61   Test Formula for Acid Thickening

Water	78.3%
Phosphoric acid (85%)	11.7%
HB	10.0%

TABLE 62 Acid Thickening Caused by Salt Addition

	% Salt added	Viscosity (cps)
Monosodium phosphate	1	10
	3	10
	5	40
	7	100
Disodium phosphate	1	10
	3	20
	5	480
	7	6,400
Trisodium phosphate	1	10
	3	160
	5	5,740
	7	17,400
Sodium chloride	1	300
	3	9,740
	5	25,200
Calcium chloride	1	80
	3	5,840
	5	11,000
	7	20,000
Potassium chloride	1	60
	3	6,300
	5	10,100
	7	16,700
Sodium sulfate	1	60
	3	460
	5	2,460
	7	5,660
Sodium acetate	1	10
	3	10
	5	60
	7	100

V. AMINE OXIDES

A. Structure and Method of Preparation

Amine oxides are well-established surfactants. The most common type is lauryl-dimethylamine oxide (AO), made from alkyldimethylamines and hydrogen peroxide (Fig. 12). Other structural features are possible. Reacting DMAPA with a

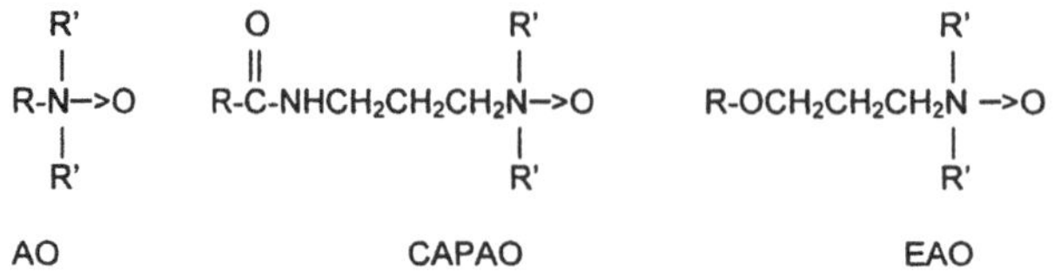

R'   O   R'   R'
| || | |
R-N->O   R-C-NHCH₂CH₂CH₂N->O   R-OCH₂CH₂CH₂N ->O
| | |
R'   R'   R'

AO   CAPAO   EAO

FIG. 12 Amine oxides. Where R = C8–22, R' is Me or $(CH_2CH_2O)nH$.

coco fatty acid or ester followed by hydrogen peroxide gives a cocoamidopropyl dimethylamine oxide (CAPAO). Both of these are typically sold as 30% aqueous solutions. An etheramine can be ethoxylated and converted to a unique amine oxide (EAO) [16].

Amine oxides are similar to betaines. In acid solutions, the amino group gets protonated and the amine oxide acts as a quat. In alkaline pH, the amine oxide behaves as a nonionic surfactant. The structure of some of the common amine oxide surfactants is given below.

B. Detergency and Other Useful Properties

Amine oxides are similar to betaines because of the protonation at acid pH they behave as cationic surfactants. At alkaline pH, amine oxides behave as nonionic. Amine oxides are good foaming agents and also act as foam booster and foam stabilizers for anionic surfactants. With anionics, amine oxides can form a complex that has better surface activity than either the anionic or the amine oxide.

Another important property of amine oxides is their resistance to oxidation, which makes them suitable for use in foaming bleaches. The bleach stability of surfactants was examined using the formulation in Table 63 and monitoring hypochlorite concentration for 6 months. The initial available chlorine was 2.0%. The test was run until the available chlorine dropped down to 1% or to 180 days, whichever came first. The results, listed in Table 64, show that amine oxides are one of the best surfactants for use with chlorine bleaches.

TABLE 63 Test Mixture for Hypochlorite Stability

Ingredient	% w/w
DI water	57.0%
Sodium hydroxide (50% solution)	2.0%
Sodium hypochlorite (5.25% solution)	40.0%
Test surfactant (100% active basis)	1.0%

TABLE 64 Hypochlorite Stability Comparison with Various Surfactants

Surfactant	Number of days until available chlorine was less than 1%	% Available chlorine after 180 days
Lauramine oxide	>180	1.25
Nonylphenol phosphate ester	>180	1.09
Sodium octane sulfonate	>180	1.16
Nonylphenol ethoxylate (9 mole)	120	
AE	110	
Cocamidopropyl betaine	2	
Sodium C_{12-15} pareth sulfate	80	
Sodium C_{14-16} olefin sulfonate	150	
Sodium dioctyl sulfosuccinate	5	

Laundry detergent, fine-fabric wash, and laundry prespotter formulations can also use amine oxides because of their good soil removal, particularly the greasy soil removal properties, and also because amine oxides act as dye transfer inhibitors for certain dyes.

The authors' work on dye transfer inhibition has shown that the commonly used nitrogen-containing surfactants like betaines, amine ethoxylates, and amine oxides all exhibit good dye transfer inhibition properties. The reason for this dye transfer inhibition is probably due to the positive charge on nitrogen which allows it to complex with the negatively charged acid and direct dyes and thus acting as a dye fixative. Table 65 lists the results of the tests ran for evaluating the dye transfer of several nitrogen-containing surfactants.

Owing to their excellent dye transfer inhibition properties, amine oxides can be used in a wide variety of laundry product formulations like laundry detergents, softergents, laundry prespotters, fine-fabric washes, etc. Amine oxides act as foam stabilizers and foam enhancers for anionic surfactants and therefore amine oxides are used in manual dishwash and shampoo formulations where foam volume, foam quality, and stability are important performance criteria. Some starting formulations using amine oxides are listed in Tables 66 through 70.

VI. OTHER SPECIALTY AMIDES

Other, even more specialized amide surfactants are known and used in selected products [17]. Sodium cocoyl sarcosinate (CS) and sodium cocoyl taurate (CT) are amide-based specialty anionic surfactants closely related to sodium cocoyl isethionate (CIS) structurally (Fig. 13) [18]. All are mild to skin and useful in personal care products. These three surfactants are some-

TABLE 65 Dye Transfer Inhibition Comparison of Various Nitrogen Surfactants

Surfactant/detergent	Dosage (active)	Dye	Delta E	Avg. delta E
Blank (no surfactant)		Red #151	31.2	
		Blue #90	47.1	
		Blue #1	19.8	32.7
AmE	1 g	Red #151	10.8	
		Blue #90	8.7	
		Blue #1	3.0	7.5
PVP	1 g	Red #151	30.7	
		Blue #90	22.4	
		Blue #1	5.1	19.4
AE	1 g	Red #151	12.7	
		Blue #90	23.5	
		Blue #1	18.5	18.2
LAS	1 g	Red #151	33.4	
		Blue #90	42.8	
		Blue #1	18.9	31.7
AO	1 g	Red #151	6.8	
		Blue #90	5.3	
		Blue #1	7.8	6.7
CAPB	1 g	Red #151	10.4	
		Blue #90	3.9	
		Blue #1	5.5	6.6
SLES	1 g	Red #151	27.4	
		Blue #90	40.4	
		Blue #1	17.1	28.3

TABLE 66 HDL Detergent with Excellent Dye Transfer Inhibition

Water	to 100%
Propylene glycol	5.0%
Enzymes	QS
Sodium citrate	5.0%
AO	10% actives
AmE	10% actives
SLES	5.0% actives
Optical brightners	QS
Dyes, preservatives	QS

TABLE 67 Fine-Fabric Wash

Water	to 100%
SLES	8.0% actives
AE	4.0% actives
CAPAO	4.0% actives
Propylene glycol	5.0%
Dyes, preservatives	QS

TABLE 68 Softergent

Water	to 100%
Imidazoline quat	8.0% actives
CAPAO	10.0% actives
AmE	10.0% actives
Propylene glycol	5.0%
Dyes, preservatives	QS

TABLE 69 Liquid Hand Dishwash Detergent

Water	to 100%
Propylene glycol	5.0%
CAPB	12.0%
SLES	8.0%
AO	5.0%
Dye, perfume, preservative	QS

TABLE 70 Hair Shampoo

Water	to 100%
Ammonium lauryl sulfate	5.0%
SLES	8.0%
CAPB	4.0%
CAPAO	3.0%
Citric acid to pH	5.5
Sodium chloride	2.5%
Dye, perfume, preservative	QS

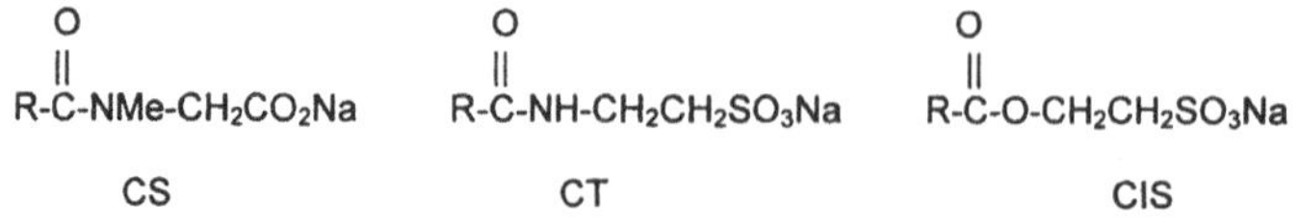

$$\text{R-}\overset{\overset{\displaystyle O}{\|}}{\text{C}}\text{-NMe-CH}_2\text{CO}_2\text{Na} \qquad \text{R-}\overset{\overset{\displaystyle O}{\|}}{\text{C}}\text{-NH-CH}_2\text{CH}_2\text{SO}_3\text{Na} \qquad \text{R-}\overset{\overset{\displaystyle O}{\|}}{\text{C}}\text{-O-CH}_2\text{CH}_2\text{SO}_3\text{Na}$$

$$\qquad \text{CS} \qquad\qquad\qquad\qquad \text{CT} \qquad\qquad\qquad\qquad \text{CIS}$$

FIG. 13   Specialty anionics.

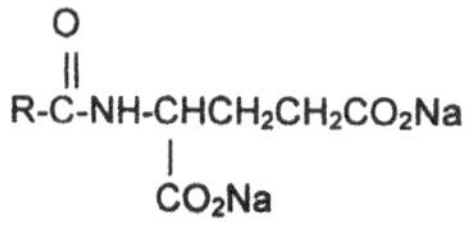

$$\text{R-}\overset{\overset{\displaystyle O}{\|}}{\text{C}}\text{-NH-CHCH}_2\text{CH}_2\text{CO}_2\text{Na}$$
$$\qquad\qquad\quad \overset{|}{\text{CO}_2\text{Na}}$$

FIG. 14   Disodium acylglutamate.

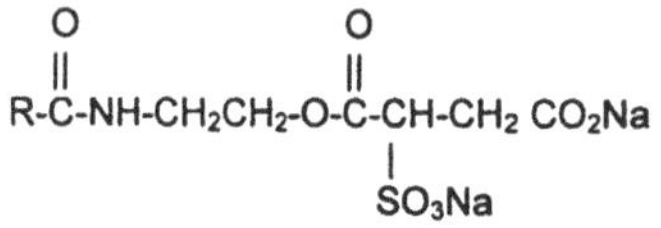

$$\text{R-}\overset{\overset{\displaystyle O}{\|}}{\text{C}}\text{-NH-CH}_2\text{CH}_2\text{-O-}\overset{\overset{\displaystyle O}{\|}}{\text{C}}\text{-CH-CH}_2\ \text{CO}_2\text{Na}$$
$$\qquad\qquad\qquad\qquad\qquad \overset{|}{\text{SO}_3\text{Na}}$$

FIG. 15   Sulfosuccinate.

$$\text{R-}\overset{\overset{\displaystyle O}{\|}}{\text{C}}\text{-O-(CH}_2\text{CH}_2\text{O)nCH}_2\text{CO}_2\text{Na} \qquad \text{R-}\overset{\overset{\displaystyle O}{\|}}{\text{C}}\text{-NH-(CH}_2\text{CH}_2\text{O)nCH}_2\text{CO}_2\text{Na}$$

FIG. 16   Carboxylated nonionics.

what difficult to make, generally requiring a fatty-acid chloride as the last step to add the alkyl portion.

Disodium acylglutamate is very similar in structure except it has two anionic groups (Fig. 14), but also requires an acid chloride in its preparation.

Sulfosuccinates of various types have long been known and used, but an alkanolamide sulfosuccinate (Fig. 15) can be prepared by reacting 1 mole of a monoethanolamide with maleic anhydride followed by sodium bisulfite.

Alkyl ethoxy carboxylates or carboxylated nonionics, as they are also called, are well known specialty nonionics which can be used as nonionics at neutral pH or as anionics at high pH [19]. Recently amide analogs of these materials have been developed [20], called amide ether carboxylates (Fig. 16). The standard method for making carboxylated nonionics is adding sodium chloroacetate to an ethoxylated alcohol, while the amide analog is made in a similar fashion from an ethoxylated monoethanolamide.

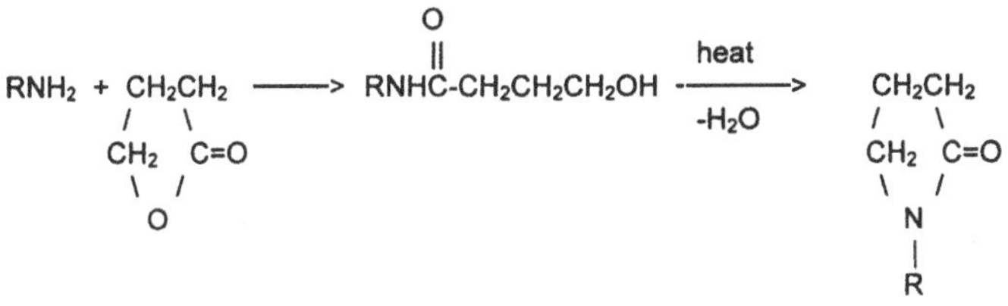

$$RNH_2 + \underset{\underset{\underset{O}{\diagdown\diagup}}{\underset{CH_2\ \ C=O}{\diagup\ \ \diagdown}}}{CH_2CH_2} \longrightarrow RNHC\text{-}CH_2CH_2CH_2OH \xrightarrow[-H_2O]{heat} \underset{\underset{\underset{\underset{R}{|}}{N}}{\underset{CH_2\ \ C=O}{\diagup\ \ \diagdown}}}{\overset{\overset{O}{\|}}{CH_2CH_2}}$$

FIG. 17 Alkyl pyrollidone surfactants.

N-alkyl pyrrolidones are a class of amide surfactants made by reacting a primary amine with butyrolactone to give a amide alcohol, then cyclizing to the pyrrolidone (Fig. 17) [21].

VII. SAFETY AND ENVIRONMENTAL ASPECTS

A. Introduction

Today, there is an increased emphasis on human safety and low environmental impact of products used. Consumers, environmental groups, and politicians, as well as surfactant manufacturers and formulators of finished products all want molecules that biodegrade quickly, have low toxicity to aquatic life, and are safe with low skin/eye irritancy to humans.

B. Biodegradation

The OECD has a thorough protocol to establish the level of biodegradability of surfactants, but the most common test is the Sturm test [22]. A number of the amphoterics and betaines are readily biodegradable (Table 71), as are both the diethanol- and monoethanolamides. Ethoxylated amines have acceptable and

TABLE 71 Biodegradation of Some Surfactants by Sturm Test

Surfactant	% CO_2 evolved (29 days)
CAPB	100
Coco AM	67
Ethoxylated amine, coco + 5 EO	60
Ethoxaylated amine, hard tallow + 2 EO	52
Ethoxylated amine, tallow + 15 EO	42
Coco diethanolamide	65
Coco monoethanolamide + 5 EO	78
Tallow monoethanolamide + 15 EO	69

complete biodegradation [23,24]; however, only structures with five or fewer EO units pass the test for ready biodegradability. Cocodiethanolamide, ethoxylated cocomonoethanolamide, and ethoxylated tallowmonoethanolamide are readily biodegradable.

C. Skin Irritation

Most of the amphoterics and betaines are have low irritation to skin and eyes [25,26]. Ethoxylated amines with a large amount of EO, such as 10 or more moles per nitrogen molecule are nonirritating to the skin. However those, analogs with 5 moles EO may be corrosive to the skin. However, many detergents with high pH from either caustic or triethanol amine are corrosive.

While DEA amides themselves are not a safety concern, residual DEA in the products is an issue, because secondary amines can oxidize to nitrosamines. Some nitrosamines are carcinogenic. For this reason, MEA amides are becoming more popular in some formulations. They can have similar chemical properties, but without the concern for nitrosamines. The MEA amides are generally nontoxic and may be slightly to moderately irritating to the skin and eyes.

VIII. SUMMARY

Specialty nitrogen-containing surfactants can offer exceptional cleaning, particularly on certain stains. Thus, when reformulating a laundry detergent or household cleaner for good cleaning and improved mildness, a betaine could be appropriate, while an ethoxylated amine could give excellent cleaning or higher concentration.

REFERENCES

1. FE Friedli, MM Watts, A Domsch, DA Tanner, RD Pifer, JF Fuller. In Proceedings of the 3rd World Conference on Detergents. Champaign, IL: AOCS Press, 1993, pp 156–159.
2. FE Friedli, MM Watts, A Domsch, P Frank, RD Pifer. In Proceedings of the World Conference on Lauric Oils. Champaign, IL: AOCS Press, 1994, pp 133–136.
3. FE Friedli, MM Watts. Chimi Oggi 13(7–8):13, 1995.
4. RA Reck. J Am Oil Chem Soc 56:796A, 1979.
5. CW Glankler. J Am Oil Chem Soc 56:802A, 1979.
6. S Arif. HAPPI February:67, 1996.
7. SH Shapiro. In ES Pattison, ed. Fatty Acids and Their Industrial Applications. New York: Marcel Dekker, 1968, pp 77–154.
8. American Society of Testing and Materials (ASTM) procedure D-3050; D-4265 and E-313 are also useful methods.
9. H Kubota, E Yamazaki, K Kodera. Jpn Kokai Tokkyo Koho JP 08,151,596 (1996).

10. American Society of Testing and Materials (ASTM) procedure D-5548.

11. FE Friedli. In J.Richmond ed. Cationic Surfactants—Organic Chemistry. Surfactant Science Series Vol. 34. New York: Marcel Dekker, 1990, pp 51–99.

12. JM Ricca, PJ Derian, F Marcenac, R Vukow, D Tracy, M Dahanayake. In Proceedings of the 4th World Surfactants Congress (CESIO), ASEPSAT, Barcelona, Spain, 1996, Vol 1, pp 302–320.

13. T Tamura, T Lihara, S Nishida, S Ohta. J Surfactants Deterg 2:207, 1999.

14. E Lomax, ed. Amphoteric Surfactants. Surfactant Science Series Vol. 59. New York: Marcel Dekker, 1996.

15. CSMA Procedure #DCC-009, Chemical Specialty Manufacturers Association Detergents Division Test Methods Compendium, 3d ed. Washington: CMSA, 1996.

16. RR Egan, MM Watts. U.S. Patent 4,263,177 (1981).

17. MJ Rosen. Surfactants and Interfacial Phenomena, 2d ed. New York: Wiley, 1989, pp 1–32.

18. K Mlyazawa, U Tamura. Cosmet Toiletries, 108:81, 1993.

19. T Cripe. U.S. Patent 5,233,087 (1993).

20. H Fukusaki, K Isobe, JK Smid. In Proceedings of the 4th World Surfactants 7 Congress (CESIO), ASEPSAT, Barcelona, Spain, 1996, Vol 1, pp 227–237.

21. RB Login. J Am Oil Chem Soc 72:759, 1995.

22. Organization for Economic Cooperation and Development. OECD Method 301 B, "Ready Biodegradability: Modified Sturm Test."

23. CG van Ginkel, CA Stroo, AGM Kroon. Tenside Surf Det 30:213, 1993.

24. P Gerike, W Jasiak. Tenside Detergents 23:300, 1986.

25. J Palicka. J Chem Tech Biotechnol 50:331, 1991.

26. Inform 4:618, 1993.

3

Sulfonated Methyl Esters

TERRI GERMAIN   Stepan Company, Northfield, Illinois

I. INTRODUCTION AND HISTORY

One of the first research organizations to evaluate sulfonated methyl esters (SMEs) was the U.S. Department of Agriculture in an effort to find uses of tallow in the mid-1950s. Stepan Company, a surfactant manufacturer in the United States, developed and sold an SME surfactant in the early 1960s, but the product was plagued by inconsistency in color and hydrolysis in finished products. Stepan, Henkel in Germany, and Lion in Japan all stepped up their research efforts to develop better processes and products after petroleum feed prices escalated in the 1970s and 1980s. These surfactants are currently used in consumer and industrial cleaners. SMEs and sulfonated fatty acids are older surfactants, but with unique properties that are very applicable to today's market.

II. STRUCTURE AND METHOD OF PREPARATION

A. Structure and Hydrolytic Stability

SMEs are prepared by the addition of sulfur trioxide to the alpha carbon of a methyl ester and subsequent neutralization with a base. Two surfactant moieties are formed (see Fig. 1). The primary surfactant formed is SME. The secondary surfactant formed is a sulfonated fatty acid. This molecule contains two different acid groups, a sulfonic acid and a carboxylic acid group. At a pH of roughly 4.5 to 9 the sulfonated fatty acid is present as a mixture of monoanion and dianion [1]. Below pH 4, the diacid is present as a monoanion. At pH < 9, the sulfonated fatty acid is present in a disodium salt form.

   The ester is stable at neutral pH, but at < pH 4 can hydrolyze to the sulfonated fatty acid. The ester at pH > 9 can again hydrolyze and form the disodium salt of sulfonated fatty acid [2]. SMEs and the various versions of the sulfonated fatty acids all have surfactant properties. These properties will be discussed later in this chapter. It is interesting to consider that, unlike amides and alkyl sulfates,

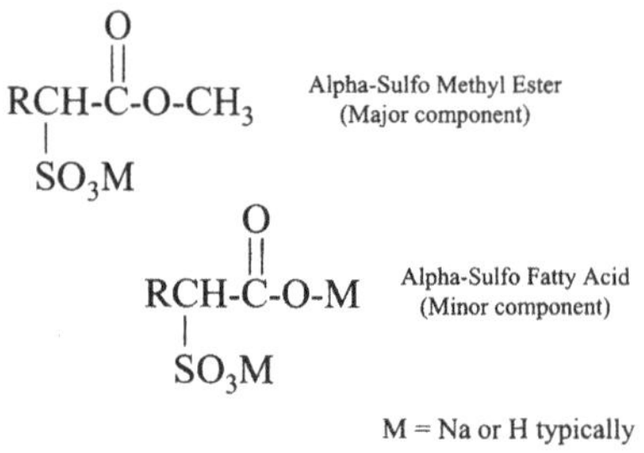

O
‖
RCH-C-O-CH₃   Alpha-Sulfo Methyl Ester
| (Major component)
SO₃M

O
‖
RCH-C-O-M   Alpha-Sulfo Fatty Acid
| (Minor component)
SO₃M

M = Na or H typically

FIG. 1   SME and sulfonated fatty acid structure.

which hydrolyze into nonsurfactant molecules, SMEs hydrolyze into other surfactants. In general, when referring to SMEs in this chapter, the sodium salt form is the indicated molecule.

B. Solubility

For analogs of the same chain length, the disalt of sulfonated fatty acid is more soluble than the SME, which is more soluble than the monoanion salt of sulfonated fatty acid at very low temperatures [3,4]. At temperatures > 10°C the SMEs become significantly more soluble because they are above their Krafft points.

C. Referenced Processes

Much has been written and presented on the manufacturing processing of SMEs and the mechanisms for forming the two moieties found in commercial products [5–11]. Of concern in the literature are the possible dark colors and high levels of the sulfonated fatty-acid species that can result. Bleaching reduces color. Both hydrogen peroxide [5,12] and sodium hypochlorite [13] have been cited as effective bleaching agents. These same references also state that bleaching can cause the SME formed to hydrolyze to sulfonated fatty acid. The neutralization process can also cause hydrolysis. To obtain the lowest colors with the highest yield of SMEs, several manufacturing processes have been suggested. Lion Corporation has developed a method of using hydrogen peroxide and methanol simultaneously [12]. The addition of methanol reesterifies the sulfonated fatty acid back to the SME. So much methanol is needed, however, that the excess methanol needs to be stripped from the product.

D. Current State of Art Processes

Satsuki wrote that 1.1 to 1.5 moles of sulfur trioxide is reacted with 1 mole of ester to form SMEs [2]. This high level of SO_3 contributed to the high colors of

SMEs made in the past. The literature also suggests long digestion times to allow rearrangement of the intermediates that are formed. Stepan Company has found that these high levels of SO$_3$ and long digestion times are not necessary to prepare commercial grade SMEs. The current milder conditions used to manufacture SMEs now also reduce the possibilities of forming undesired by-products. These byproducts will be discussed in the Safety and Environmental portion of this chapter.

III. PHYSICAL CHEMISTRY

SMEs and sulfonated fatty acids are anionic surfactants. This section will highlight the effects of carbon chain length, counter ions, and combinations with other surfactants.

A. Surface Tension

Both SMEs and sulfonated fatty acids are effective at reducing surface tension and are similar to other anionic surfactants such as alkylsulfates and alkylbenzenesulfonates. As expected with most anionic surfactants, as the carbon chain on the SME or sulfonated fatty acid increases, the critical micelle concentration (CMC) decreases [3]. In Figure 2 it can be seen that the SME has a lower CMC than the corresponding disodium sulfonated fatty acid [14]. Table 1 shows that the SME and its corresponding sulfonated fatty acid have very similar CMCs [15, 16]. Surprisingly, as shown in Table 2, increasing chain length has little effect on surface tension measurements of SMEs and various salts of sulfonated fatty acids [4, 15]. The monosodium salts of sulfonated fatty acids (lauryl and myristyl) exhibit the lowest surface tension compared to the corresponding SME, diacid, and disodium salt of the same chain length. The disodium salts

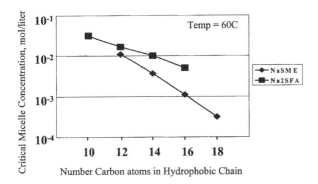

FIG. 2   Figure 4 of Henkel-Referate. CMC versus number of Carbon atoms.

TABLE 1 Critical Micelle Concentrations

$C_nH_{n+1}CHSO_3XCOOM$ where n =	CMC, % active where X = Na, M = CH3	CMC, % active where X and M = H
10	N/A	0.36
12	0.09	0.07
14	0.014	0.02
16	0.003	0.005

Source: Refs. 15, 16.

TABLE 2 Surface Tension of Surfactant in Deionized Water, dynes/cm

$C_nH_{n+1}CHSO_3XCOOM$ where n =	Surface tension where M = CH$_3$, X = Na (sodium sulfonated methylester)	Surface tension where M = H, X = H (diacid of sulfonated fatty acid)	Surface tension where M = H, X = Na (monosodium salt of sulfonated fatty acid)	Surface tension where M = Na, X = Na (disodium salt of sulfonated fatty acid)
10		40.9@0.1%	32.7@0.1%	69.3@0.1%
12	39.9@0.2%	37.1@0.1%	30.2@0.1%	62.8@0.1%
14	40.2@0.2%	41.5@0.1%		44.3@0.1%
16	39.0@0.2%	43.5@0.1%		

Source: Refs. 4, 15.

have very high surface tensions at the concentrations tested. These results are not surprising in light of the fact that these molecules at 0.1% aqueous 25°C are still significantly below their CMC concentrations [3].

B. Micelle Size and Solubilization

A determination of micelle size can give an indication of how well a surfactant can solubilize oils. The larger the micelle, theoretically, the more soil that can be incorporated inside the micelle and therefore the better the cleaning. Satsuki stated in his paper that linear alkylbenezenesulfonate (LAS) and SME micelles are similar in size and are approximately twice as big as sodium lauryl sulfate (SLS) [17]. The average number of molecules per micelle, the aggregation number, is twice as high for SME as for the LAS. The aggregation number for the SLS is roughly 2.5 times that of LAS. His solubilization work showed that the densely packed SLS micelles and the loosely packed LAS micelles were not able to increase in size compared to the SME as more polar oil was added to a surfactant solution [3].

Kamesh Rao determined the effects of counterions on SMEs [18]. Magnesium SME has larger micelles than the sodium or ammonium salts, 5 nm versus 2 nm and 3 nm, respectively. Two performance tests that are dependent on the amount of oil or grease that can be solubilized and emulsified support the hypothesis that larger micelles can improve cleaning. A modified Colgate Mini-Plate Test and the Baumgartner grease removal test both showed that the Mg SME outperformed its Na and NH$_4$ salt analogs [18].

IV. PERFORMANCE

Physical-chemical properties do not always predict how a surfactant will perform in use. The measurement of surfactant properties such as foaming, wetting, emulsification, coupling, etc., used in combination with the physical chemical properties of surfactants can give a better understanding of how a surfactant could or does behave in final commercial uses.

A. Surfactant Properties

1. Foaming

Many factors can affect foam. There are also many ways to measure foam such as the Ross-Miles Test, the Shake Foam Test, and the Blender Foam Test to name a few. When studying foam, generally the following characteristics are evaluated: initial foam height or flash foam, foam stability over time, solution drainage, and foam morphology. The factors that can affect foam include concentration of test material, water hardness, water temperature, and presence of soil. To discuss the foaming properties of SMEs and sulfonated fatty acids, included here are references to data generated via the Ross-Miles Test [1,4,19,20].

An analysis of the data available shows that the foaming ability of SME increases as chain length increases, reaching a maximum at approximately 14 to 16 carbons depending on temperature. As the chain length exceeds 14 carbons the flash foam generated begins to decrease. Figure 3 demonstrates not only the effect of carbon chain length but also the effects of water hardness [20]. Unlike LAS and alkyl sulfate surfactants, as water hardness increases, the foam properties of SMEs are enhanced. It is not until the carbon length is C18 and water hardness exceeds 300 ppm as CaCO$_3$ in the test solution that foam is adversely affected. Figure 4 also shows an increase in foam height as the carbon chain of the mono and di salts of sulfonated fatty acid is increased, peaking at 16 carbons [1,4]. In order to solubilize the sulfonated fatty acids, the test was conducted at 60°C.

Figure 5 shows the effects of monovalent and a divalent cation on the initial foam height of a C14-SME [20]. Again, the effects of water hardness are also shown in the same graph. In soft water, the magnesium and ammonium salts

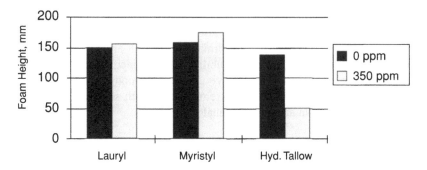

FIG. 3   SME foam data chain length and water hardness.

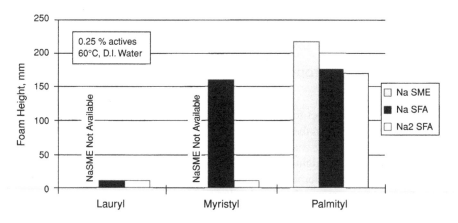

FIG. 4   Foam data on SME and sulfonated fatty acid, varying chain length.

gave the highest foam. In hard water, the sodium and ammonium salts gave the highest foam. An increase in water hardness results in higher flash foam for all except the magnesium salt. It may be that the higher foam is a result of interactions between the divalent cations and the SME. Since the Mg salt brings with it its own divalent cation, there is no improvement in foam when hard water is used.

The effect of counter ions is much more dramatic with the sulfonated fatty acid than with the SME. J.K. Weil et al. performed foam tests on various mono and di salts of a C16-sulfonated fatty acid and a C18-sulfonated fatty acid. In Figure 6 few trends can be seen [21]. The half salts of the magnesium and calcium C 16-sulfonated fatty acids foam higher than the half-salts of the monovalent analogs. Yet, this trend is not true for the C18-sulfonated fatty acid. The data

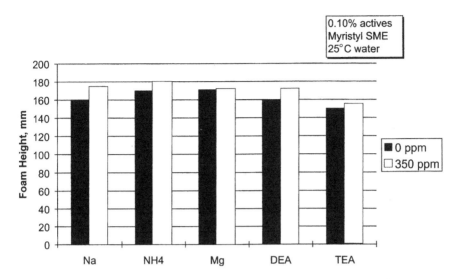

FIG. 5 Foam data on SME, varying counterion.

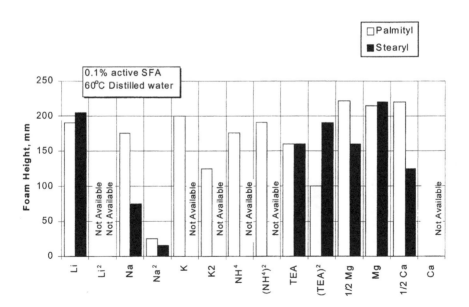

FIG. 6 Foam data on sulfonated fatty acid, varying counterion.

available show that the half-salt of the lithium C18-sulfonated fatty acid and the full salt of the magnesium C18-sulfonated fatty-acid foam the highest. Regardless, with a few exceptions, all the salts of the sulfonated fatty acids studied are high-foaming surfactants (foam heights > 120 mm).

Surfactants are rarely used as single components in formulations. Combinations of surfactants are generally used to obtain maximum performance and desired physical characteristics. Blending surfactants can result in enhanced performance over that of the individual components. SMEs demonstrate this synergy phenomenon when blended with other anionic surfactants. In Figure 7, a foam synergy is shown between blends of a commercial product available from Stepan Company, ALPHA-STEP MC-48, and a sodium lauryl ethoxysulfate containing an average of 3 moles of ethylene oxide units [22]. The commercial surfactant is a blend of sodium SME and monosodium-sulfonated fatty acid at a ratio of about 5:1. The average carbon chain length of both moieties is 13.6 coming from a stripped coco chain distribution.

2. Wetting

Another important property of surfactants is their ability to wet surfaces. Performance data in the literature used the Draves Wetting Test [23] and the Binding Tape Test [24]. SMEs and the mono- and disodium salts of sulfonated fatty acids show opposite behavior as chain length increases. As the chain length increases from 14 carbons to 18 carbons, SME wetting times increase, indicating that longer chain lengths are detrimental to wetting ability [15]. The opposite is observed for sulfonated fatty acids [4,15] (Figs. 8, 9). It is generally believed that

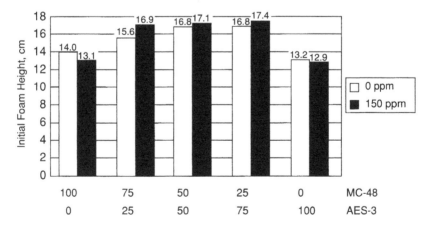

FIG. 7   Foam data on SME/AES synergy.

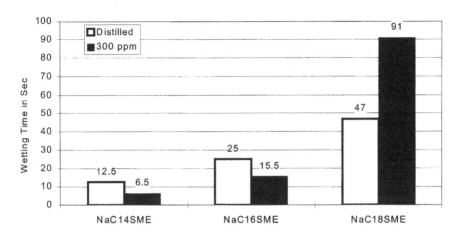

FIG. 8   Wetting data of SME.

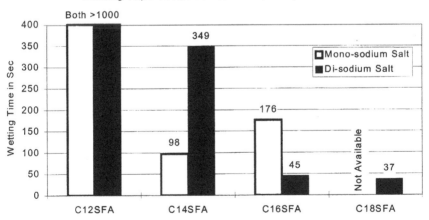

FIG. 9   Wetting data of sulfonated fatty acid.

SMEs are better wetters than the mono salts of sulfonated fatty acids [4] (Fig. 10). Different counterions to the sulfonated fatty acid can have a significant effect on wetting, but the SME analog is a better wetter, at least for the C16 versions.

Just as with foaming properties, blends of surfactants can show synergistic wetting behavior. Blending a C14-SME with coco soap and with SLS gives much improved wetting times [21] (Fig. 11). Likely, there are certainly more combinations of other surfactants with SMEs than can result in improved wetting behavior.

3. Coupling

Drozd mentions that SMEs are good solubilizers [25]. Smith describes SMEs as hydrotropes [26]. Stein mentions that SMEs can reduce the viscosity of surfactant solutions [5]. Viscosity reduction, solubilization, and cloud point reduction are all properties of hydrotropes. Surfactants that function as hydrotropes are generally short-chained molecules such as sodium xylene sulfonate (SXS) and sodium octane sulfonate. These hydrotropes do not demonstrate high foam or detergency properties. Surfactants that accomplish more than one function in a formulation can be a benefit. It is generally more desirable to use one surfactant for several properties than to use a detergent surfactant and a separate hydrotropic surfactant to keep a formulation clear and homogeneous. Shipping and storing a minimum number of ingredients can have economic advantages.

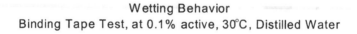

Wetting Behavior
Binding Tape Test, at 0.1% active, 30°C, Distilled Water

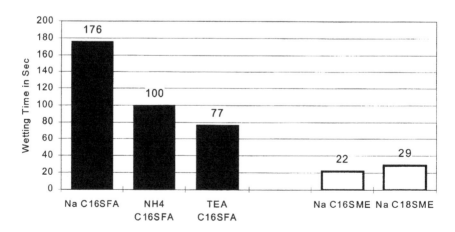

FIG. 10   Wetting data, different counterions.

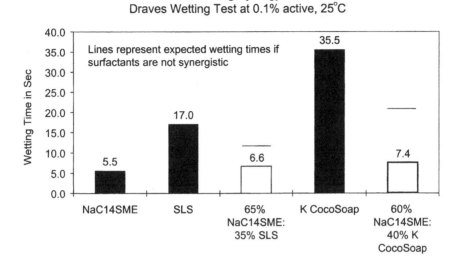

FIG. 11   Wetting synergy, SME + SLS/soap.

Malik presented a paper on multifunctional surfactants [27]. Included were commercial-grade blends of C12 SME/sulfonated fatty acid and C13.6 SME/sulfonated fatty acid. These products were compared to traditional industry hydrotropes as well as to other specialty surfactants. The C12 and C13.6 SME/acids are very effective hydrotropes as compared to SXS [28] (Fig. 12). Later in this chapter the detergent and hydrotropic properties of SMEs will be used in formulations.

4.   Viscosity Modification

As mentioned earlier, hydrotropes can reduce viscosity. This is also a property of SMEs. In Figure 13 the viscosity versus sodium chloride response curve of a test formulation is shown. Different commercial grade anionic surfactants are evaluated for their viscosity response. The NaLAS was a C11.4 high-2-phenyl, high-tetralin sodium linear alkylbenzenesulfonate; NaAES was a sodium 3-mole ethoxylaurylsulfate; Na SME was a blend of C13.6 SME/sulfonated fatty acid; and the NaAOS was a sodium C14, 16 alpha-olefinsulfonate.

As known in the industry and shown again in Figure 13, alcohol sulfates typically give the highest viscosity values when blended with a viscosity modifier such as the 1:1 cocodiethanolamide used here; 1:1 refers to the mole ratio of diethanolamine to fatty acid used to prepare the alkanolamide. The Figure 13 curves end where the next addition of sodium chloride would cause the test solution to "salt out." What is seen is the ability of SMEs to maintain a low viscosity

Hydrotropic Properties

Ingredient	Wt %
Nonylphenol Ethoxylate, 9 Mole EO	5.0
Nonylphenol Ethoxylate, 4 mole EO	5.0
Trisodium Citrate	1.0
Hydrotrope Requirement	q. s.
Water, Deionized	Balance

Hydrotrope	Wt. of Actives (g)*
C12 SME/SFA Blend	1.2
C13.6 SME/SFA Blend	1.5
SXS (Sodium Xylene Sulfonate)	4.4

FIG. 12   Coupling properties of SME. *Gram weight of actives required to clarify 100 grams of base formulation at 26°C.

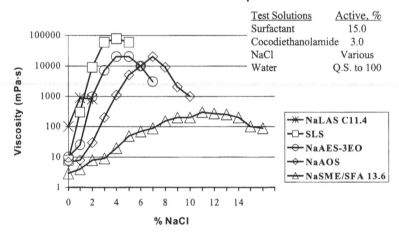

FIG. 13   Viscosity vs. NaCl curves.

compared to other anionic surfactants as well as their ability to maintain a clear solution in the presence of high levels of an electrolyte.

5.   Water Hardness Tolerance and Metallic Ion Stability

Some surfactants, such as SLS, are affected by hard water. What this means is that the di- and trivalent counterions (Ca, Mg, Fe, etc.) replace the sodium ion of

TABLE 3 Lime Soap Dispersing Properties of Anionic
Surfactants Including Tallow SME

Surfactant	% Surfactant needed
Sodium tallow SME	**22**
Sodium lauryl sulfate	>100
Na Lauryl ethoxysulfate—3 mole	22
Na Lauryl ethoxysulfate—12 mole	39
Na Lauryl ethoxysulfate—30 mole	>100
Na linear alkylbenzene sulfonate—C11.4	>100
Na alpha olefin sulfonate—C14, C16	>100

Source: Ref. 29.

SLS and cause the surfactant to become significantly less water soluble. Surfactant solutions will typically become hazy and over time the surfactant can precipitate. If a surfactant precipitates, it is not available for foam generation, cleaning, wetting, etc. These surfactant properties require a surfactant be available at an interface, not sitting on the bottom of a container. SMEs are not affected by hard water. In other words, their divalent salts are water soluble. Sulfonated fatty acids are affected by hard water and can precipitate in cold water [3]. The effects of water hardness on detergency will be discussed later.

Stirton measured the stability of C16,18 SMEs to multivalent ions [4]. He found that these SMEs are very stable to magnesium, calcium, iron, nickel, copper, and zinc. They were least soluble in the presence of aluminum and lead.

6. Lime Soap Dispersing

One of the often noted properties of SMEs is their lime soap dispersing properties. Lime soap, or the calcium salts of long-chain fatty acids, collects in showers and tubs as soap scum. In laundry, lime soap built up on fabric over time can cause graying. Yet, soap is an inexpensive surfactant and its incorporation into formulations can be very beneficial. Using both a lime soap dispersing surfactant and soap can overcome the disadvantages of using soap alone.

Knaggs determined that the C16 SME had the best lime soap dispersing power of the chain lengths tested (C12 through C18) [20]. He used the Borghetty test [28] modified to use sodium 80/20 tallow/coco soap instead of sodium oleate, hard water of 1075 ppm as $CaCO_3$ instead of 1000 ppm, a Ca:Mg ratio of 63:37 instead of 60:40, and test surfactant concentrations equal to 0.50% active instead of 0.25% active. All changes were considered minor except for the soaps used. Knaggs found that the C12 SME is a poor lime soap dispersant. The C14, C16, and C18 versions are as good or better than sodium N-methyl N-tallow acid taurate, a known lime soap dispersant.

In Table 3, sodium tallow SME demonstrates very high lime soap dispersing properties compared to other commodity anionic surfactants [29]. The same modified Borghetty test as mentioned above was used. The lower the percent surfactant needed to disperse the calcium soap, the more effective the surfactant. It must be noted that the surfactants used were commercial grade and are not pure cuts.

V. USES

Surfactants are used for a variety of reasons—foaming, wetting, emulsification, etc. One of their primary functions is the removal of soil from surfaces. In this section, the detergency properties of SMEs and sulfonated fatty acids will be discussed. Included will be suggested or actual uses of the surfactants in commercial cleaners.

A. Laundry

Much has been written on SME's and sulfonated fatty acid's potential in laundry applications. So much so that the following only summarizes some of the many studies that have been done.

Satsuki studied sodium SME of different chain lengths in test solutions also containing sodium carbonate and silicate [2]. He included LAS and SLS for comparison. In 25°C, 54-ppm hardness water, the surfactants are ranked thus for detergency: C16 SME > C18 SME > LAS > C14 SME > SLS > C12 SME. In harder and warmer water, 40°C, 270-ppm as $CaCO_3$, a different ranking was found: C16 SME = C18 SME > C14 SME > SLS > C12 SME > LAS. Overall, the C16 SME and C18 SME exhibited the highest detergency. All test solutions performed better in 25°C, 54-ppm water than in the 40°C, 270-ppm water.

Satsuki also studied the effects of surfactant concentrations on detergency in combination with sodium carbonate and silicate. He found that less C14,16 SME is required to obtain the same level of detergency compared to LAS [2].

Satsuki then examined the effects of several anionic surfactants versus water hardness in phosphate-built formulations. The SME outperformed the alpha-olefin sulfonate, LAS, and SLS tested at all water hardnesses other than 0 ppm. It is believed that SME's tolerance to water hardness ions allows for higher detergency [2].

Rockwell and Rao show data comparing C12 SME and C13.6 SME to LAS and SLS via detergency testing [30]. In the article they explain that these are commercial products and not pure SME samples. They do contain some sulfonated fatty acid as an active surfactant. The testing was accomplished in a Terg-O-Tometer with 1 L of wash liquor and the delta reflectance represents the difference in reflectance units of the "L" reading of the "Lab" scale via a Hunter Colorimeter. Rockwell demonstrated poorer performance of the C12 SME in soft water compared to the other surfactants. In hard water, the SLS and the 1:1 combination of SLS and C13.6 SME were most effective on a 65/35 polyester/cotton blend. On 100% cotton under cold-water conditions, both the SMEs outperformed the other surfactants in hard water.

Satsuki again compared various-chain-length SMEs to LAS and SLS in an experimental system built with sodium carbonate and sodium silicate [17]. He specifically stayed away from sequestering builders, e.g. phosphates. He evaluated the surfactants in 25°C, 54 ppm as $CaCO_3$ water as well as in 40°C, 270 ppm as $CaCO_3$ water. The fabric was cotton and the soil an artificial soil explained in the article. His studies revealed that the C16 SME gave the highest detergency, with the C18 SME close behind under both conditions. Further experimentation by Satsuki showed that 200 ppm C16-SME could equal or outperform 300 ppm of either the LAS or the SLS at 25°C, 54-ppm water hardness. Similar results were obtained with natural facial soil.

Stein and Baumann studied the detergency of tallow SME and palm kernel SME [5]. The testing took place in a Launder-ometer at 90°C on cotton and at 60°C on a cotton/polyester blend. Water hardness was 300 ppm as $CaCO_3$. They found that without a builder present, the tallow SME outperformed the palm kernel SME, which outperformed the LAS also included in the study. It was not until a nonionic surfactant and tripolyphosphate were added to the anionic surfactants that equivalent detergency was obtained. Stein and Baumann cite the low sensitivity to water hardness as the reason for the high detergency of the SMEs.

Ahmad and colleagues studied SMEs made from palm stearin and palm fatty-acid distillates [8]. They conducted Terg-O-Tometer studies of built laundry powders, with and without phosphate. Evaluations took place in water hardness conditions ranging from 50 ppm to 500 ppm as $CaCO_3$ and from water temperatures ranging from room temperature to 60°C. They concluded that there was no difference in performance between the SME made from palm stearin and palm fatty-acid distillate. The SMEs were equal to or better than LAS in the nonphosphate test formulation at all conditions. The SMEs were equal to or better than LAS in both test formulations, with or without phosphate, in soft water.

Schambil studied detergency effects of chain length on SME and the disodium salts of sulfonated fatty acid [3]. He found that as chain length increases, detergency improves in both soft and hard water. The highest performers in both surfactant classes were the C16 and C18 chain lengths. The sulfonated fatty acid analogs did not perform as well as their SME counterparts. Increasing water hardness decreased the performance of both the SMEs and sulfonated fatty acids similarly for the C14 to C18 samples. Only the C12 SME had better detergency in hard water than in soft water.

If Zeolite was added, the hard water performance of the SMEs increased to equal or better performance compared to soft water conditions without the Zeolite. The sulfonated fatty-acid samples in hard water with Zeolite showed approximately 85% of the remission values of the analog SME samples in soft water. The addition of a builder significantly improved sulfonated fatty-acid performance. The Zeolite is necessary for sulfonated fatty-acid detergency performance due to their sensitivity to water hardness.

Stirton studied the detergency properties of sulfonated fatty acids [4]. He and his colleagues found that the di-Na C18-sulfonated fatty acid was the best detergent. The more soluble salts, i.e. the mono- and disodium salts of C12-sulfonated fatty acid and the disodium salt C14-sulfonated fatty acid, gave the poorest detergency. Changes in chain length had a larger impact on detergency than changes in water hardness.

Weil and his colleagues studied different cations of mono and di salts of palmityl and stearyl sulfonated fatty acid. They found that the monolithium, triethanolammonium, magnesium, and calcium salts gave the highest detergency [21].

Stirton compared the disodium salt of C16-sulfonated fatty acid to SLS and LAS [31]. He concluded that the disodium C16-sulfonated fatty acid is a good detergent in both soft and hard water. His results show that the sulfonated fatty acid statistically outperformed the SLS and alkylaryl sulfonate in his particular test (Fig. 14) [31].

The detergency of differing ratios of the C13.6 analogs of SME to sulfonated fatty acid were evaluated using a Terg-O-Tometer with cotton, 35/65

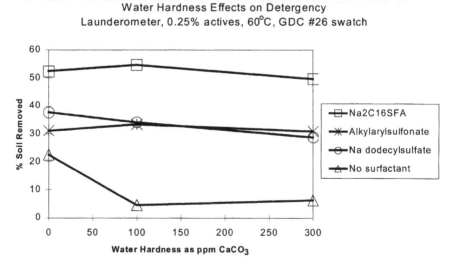

FIG. 14   Detergency of C16 sulfonated fatty acid, SLS, LAS vs. water hardness.

cotton/polyester, and polyester fabric swatches soiled with dust/sebum soil from Scientific Services in New Jersey [32]. Figures 15 and 16 show the detergency results on cotton and polyester, respectively. Twenty-five percent of the total surfactant can be the sulfonated fatty acid and not diminish detergency. Detergency trends were similar on the cotton/polyester blend. Detergency improved with high water hardness on polyester. As expected from the literature, increasing water hardness decreased the detergency of the 100% sulfonated fatty acid.

R.G. Bristline Jr. and colleagues studied the detergency properties of tallow SME with tallow soap, with and without builders [33]. Testing was conducted in a Terg-O-Tometer with various soils on cotton swatches. The detergency of soap was significantly enhanced by the incorporation of the tallow SME and builders such as sodium tripolyphosphate, sodium citrate, sodium metasilicate, and sodium oxydiacetate. Sodium carbonate only affected detergency results in multiwash tests. Tallow SME was found to be as effective at improving detergency as other known lime soap dispersing agents, e.g., tallow alcohol ethoxysulfate and ethoxylated n-hexadecanol-10 ethylene oxide units (EO). LAS was also studied and performed poorer in comparison to the other lime soap dispersing agents (LSDA) including tallow SME. Optimum detergency was obtained in most cases with an 80:10:10 ratio of soap:LSDA:builder. Test conditions: 300 ppm water, 60°C, and 0.2% surfactant plus builder. Detergency was measured by delta reflectance.

Overall the C16 and C18 chain lengths of both SMEs and sulfonated fatty

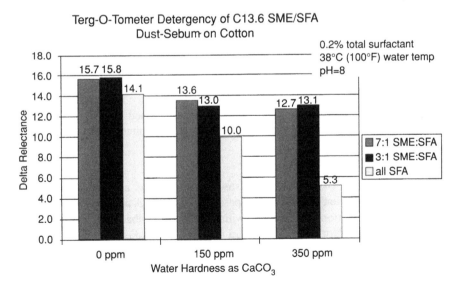

FIG. 15 SME/sulfonated fatty acid Terg at pH 8 on cotton.

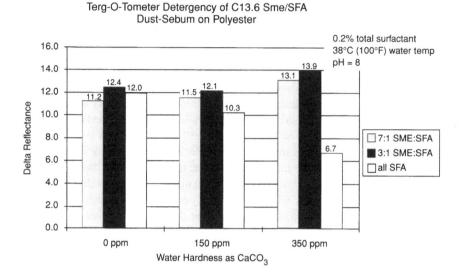

FIG. 16   SME/sulfonated fatty acid Terg at pH 8 on polyester.

acids are the best detergents of these surfactant classes. The SMEs generally out-performed LAS and SLS in hard-water conditions. In built or unbuilt formula-tions the SMEs are good detergents. The data suggest that although the sulfonated fatty acids are not as detersive as the SMEs, they are still very good surfactants. In an optimized formulation they should give acceptable perfor-mance. With their lower solubilities, they may find more applicability in pow-dered, bar, or paste formulations.

Long-chain SMEs, C14 through C18, are used in commercially available powdered laundry products in Japan. The shorter-chained SMEs, C12 through C14, have found use in liquid laundry products in the United States.

B.   Hand Dishwash

Light-duty liquids, of which hand dishwashing products are the major subset, have seen formulation changes in the past several years. A previously static cate-gory, the introduction of products with claims of antibacterial activity have spurred growth in North America. Ultra products have taken over the U.S. mar-ket. Microemulsion technology was introduced in Europe. Mildness and grease cutting have become key attributes of hand dishwash products. Surfactants pro-viding more than one property will have the advantage.

SMEs' high foaming and hydrotropic properties make them candidate surfactants for liquid hand dishwash products. This type of product can be from 10% to 45% total active surfactant. Foam actually is a more important attribute than detergency, although a claim to cut through grease on dishes helps a product's marketability.

Drozd showed that substituting a C13.6 SME for a lauryl ethoxy(3)sulfate in a liquid dishwash formulation resulted in equal performance via the Colgate Mini-Plate Test [25,34]. Also, less hydrotrope was required to obtain optimum physical properties. Substitution with the C12-SME resulted in lower performance.

One study in France showed the foam and cleaning synergy between a C13.6 SME/sulfonated fatty-acid blend and SLS [35]. In Figure 17 it can be seen that as SME replaces from 15% to 46% of the SLS actives, the number of washed plates before a foam endpoint is reached is at its highest. The test method used was a large plate test. The soil consisted of water, powdered milk, potato flakes, and hydrogenated coconut oil; 4 mL soil per plate. The test solution was 5 L of 0.032% actives in 220 ppm water as $CaCO_3$ at 45°C.

According to Rao and associates, the addition of magnesium ions can greatly improve the dishwash performance of SMEs [18]. This is not surprising in light of the previously mentioned improved foam properties and micelle size of Mg-SME versus Na-SME. He showed that Mg-SME washed 27

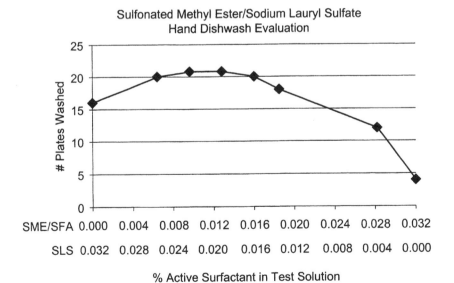

FIG. 17 SME/SLS large-plate test for synergy.

miniplates whereas the sodium and ammonium analogs washed only 13 and 12 miniplates, respectively. His test solutions were 0.1% active in deionized water.

Using alkyl ether sulfate and SMEs together in a MgLAS/amide based formulation can result in increased performance [36]. A formulation containing both ether sulfate and SME washed 51 miniplates, whereas the equivalent formulation, containing only the ethersulfate, washed 45 miniplates.

Sajic and colleagues reported that the same SME/ether sulfate formulation in the above patent could outperform leading commercial U.S. hand dishwash products in the removal of lard from plastic [37]. The SME formulation was either equal to or better than the commercial products via the miniplate test.

Knaggs and colleagues published data on the use of the sodium and magnesium salts of myristic SME in hand dishwash evaluations [20]. Using the Mayhew test [38], Knaggs concluded that when formulated with the proper foam boosters, SMEs allow for the formulation of products that performed equal to commercial products at 50% to 60% of their concentration.

C. Hard-Surface Cleaners

One of the most diverse consumer and industrial product categories is hard-surface cleaning. These products can contain from < 1% surfactant, as found in glass cleaners, to > 70% surfactants, as found in some concentrated industrial cleaners. They can be highly acidic toilet bowl cleaners or highly alkaline floor wax strippers. Unlike dishwash and laundry products, surfactants may not be the majority of the composition. They are certainly key ingredients. While some work has been published on the use of SMEs in hard-surface cleaners, no references on use of sulfonated fatty acids in hard surface were found.

Malik presented data [27] on the detergency of C12 and C13.6 SMEs via a modified Gardner Straight Line Washability Test [39]. The test solutions consisted of 0.6% of the test surfactant and 0.12% tetrapotassium pyrophosphate in 140 ppm as $CaCO_3$ water at 25°C. The C12 SME gave equivalent performance compared to decyl and branched dodecyl diphenyl oxide disulfonates. The C13.6 SME slightly outperformed the C12 SME but was not quite as good as the C8–10, 5-mole ethoxylated phosphate ester that was also included. Both SMEs contained the analog sulfonated fatty-acid sodium salts as secondary surfactants.

One study compared the performance of the commercially available C13.6 SME/sulfonated fatty acid blend to a 3-mole sodium lauryl ether sulfate and to NaLAS with an average carbon chain of 11.4 [40]. Again, using the modified Gardner test, it can be seen in Figure 18 that the LAS and SME/sulfonated fatty acid blend gave equal performance. The ether sulfate was slightly infe-

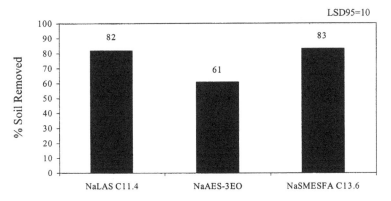

FIG. 18   LAS, AES, SME Gardner HSC data.

rior. The test solution was 2% active surfactant in deionized water without builders present.

Combining the SME performance data with the previously mentioned hydrotropic evaluations suggests they may be good choices as hydrotropes in hard-surface cleaners. The highly acidic and alkaline formulations will need to be avoided due to SME's tendency to hydrolyze in extreme conditions. Formulations that contain high amounts of surfactant where high foam is desirable are also good candidate formulations for SMEs.

D. Soap Bars

One application for SMEs and sulfonated fatty acids not usually found in the literature is soap bars. These can be personal wash/toilet bars or laundry bars. A 1942 patent discusses the use of SMEs in soap bars [41]. SMEs are also typically included in the extended list of anionic surfactants in more recent bar soap patents. However, they are rarely the focus of the patent. Sulfonated fatty acids do promote "feel" and improve "smear" properties [42,43]. More recently, the use of SMEs and sulfonated fatty acids in specific ratios was recommended [44].

There appears to be a synergy between SMEs and sulfonated fatty acids, which allow for improved foam and feel of a soap bar while retaining ease of production. Existing equipment used to make pure soap bars may be able to make combo SME/sulfonated fatty acid/soap bars. Surfactants currently used in combo soap bars include alkylether sulfates, sodium cocoylisethionates, alkyl glyceryl ether sulfonates, and betaines. Incorporation of these surfactants generally requires special equipment and processing.

VI. SAFETY AND ENVIRONMENTAL

A. Biodegradation

SMEs are easily biodegraded under aerobic conditions. E.W. Maurer et al. found that as the alkyl chain increases from 9 carbons to 18 carbons, percent biodegradation increases [45]. Using an Esso controlled nutrient procedure, the activated sludge used was acclimated to the test samples and biodegradation was measured via loss of carbon. This same article also shows that the lauryl analogs of SME and sulfonated fatty acid are equivalent in biodegradability. Drozd published primary degradation data on C12 and C14 SMEs [25]. The results show a reduction of surfactant actives via Methylene Blue titration evaluations. After 8 days, >99% of the actives had proceeded through primary degradation.

Mitsuteru Masuda et al. studied the biodegradation pathways of a C14 SME [46]. They found that the microbial attack began with ω-oxidation to form a carboxyl group and then continued with β-oxidation, removing 2 carbons at a time. A temporary intermediary monomethyl α-sulfosuccinate was formed which then underwent desulfonation.

Unfortunately like LAS, there are reports that SMEs do not degrade quickly under anaerobic conditions. Anaerobic degradation is important if a compound does not degrade during sewage treatment and is therefore present in a significant amount in the resulting "sludge." Sewage sludge is sometimes put into landfills and is also used as fertilizer. LAS has not been found to be an environmental concern via numerous studies of treatment plants and waterways [47–52]. There is no reason that SMEs and sulfonated fatty acids should be considered any less environmentally acceptable.

B. Toxicity

Toxicity to people and the environment needs to be considered when using any chemical or product. Following is some information on the safety of SMEs and sulfonated fatty acids.

1. Acute

A 36% to 38% active C12 SME was found to be nontoxic orally. The LD_{50} is > 5 g/kg [25].

2. Aquatic

Aquatic toxicity of SMEs is significantly affected by the chain length of the alkyl group. As the chain length increases, the toxicity increases. Drozd published reported LC_{50} values of various surfactants tested on fish and invertebrates [25]. The C12 SME data ratings are "practically nontoxic" according to the U.S. Fish and Wildlife Service. Other surfactants listed, such as LAS, AOS, and alcohol ethoxylates, are in the range of "moderately toxic." Henkel published data on fish and invertebrate toxicity of C16/C18 SMEs [53]. The fish toxicity, LC_o data, of the C16/18 SME was 0.4 to 0.9 mg/L. The C12/14 SME data showed significantly less toxicity at 46 mg/L.

C. Mildness

SMEs are recognized as mild surfactants. The C12 SME was compared to a lauryl 3-mole ether sulfate and to a lauryl 3-mole ethoxysulfosuccinate [25]. At 10% actives the SME was as mild as the sulfosuccinate and less irritating than the ether sulfate. Table 4 shows mildness and toxicity data on commercial products of SME/sulfonated fatty-acid blends [54]. An increase in alkyl chain length resulted in an increase in aquatic toxicity and in rabbit skin irritation. Human skin irritation studies did not show the same trend.

A 14-day cumulative irritation test was run [40]. The number of days it takes a product to produce irritation (score of 1) were determined. The longer it takes a product to reach a score of 1, the milder the product. Table 5 shows a comparison of various surfactants to sodium lauryl sulfate, an irritating control. As can be seen, the SME and sulfonated fatty-acid samples evaluated were as mild as the 3-mole ether sulfate and isethionate tested. These surfactants are recognized in the industry as mild.

TABLE 4 Irritation and Toxicity Data of Commercial SME/Sulfonated Fatty-Acid Blends

Evaluation	Alpha-step ML-40 C12 SME/sulfonated fatty acid	Alphpa-step MC-48 C13.8 SME/sulfonated fatty acid
Human patch test for mildness (0.25%, 0.50%, 0.75% actives)	Mild	Mild
Primary skin irritation, rabbit (10% active)	Slightly irritating	Severely irritating (similar to AES-3EO)
Primary eye irritation, rabbit (10% active)	moderately irritating	Moderately irritating
Acute oral toxicity, rat	LD_{50} >5 g/kg	LD_{50} >5 g/kg
Acute aquatic toxicity, rainbow trout	LC_{50} = 129 mg/L	LC_{50} = 4.7 mg/L
Acute aquatic toxicity, *Daphnia magna*	EC_{50} = 200 mg/L	EC_{50} = 11.8 mg/L

Source: Ref. 54.

TABLE 5   Comparative Irritation Study via 14-Day Cumulative Irritation Test

Product	Average days to produce irritation
Sodium lauryl sulfate	2.70
Sodium cocoylisethionate	4.27
C 13.6 sulfonated fatty acid	4.97
Sodium lauryl ether sulfate—3 moles EO	5.07
C13.6 SME/sulfonated fatty acid 7:1 ratio	5.20

D. Sensitization

There have been published concerns of possible impurities in commercial production of SMEs and sulfonated fatty acids. Of specific concern are dimethylsulfate (DMS), dimethylsulfoalkanoate (DSA), isoestersulfonate (IES) [54], and internal unsaturated sultones (IUS) [55].

Despite D.W. Roberts et al.' concerns over DMS, DSA, and IES sensitization and toxicity potentials (real or proposed), they indicate in their paper [54] that it is unlikely any of these molecules would survive through to the finished neutralized product. Analytical data support that DSA and DMS are not found in commercial products [40]. Even in Roberts' study, if methanol is present prior to neutralization, IES was not detected. It is well known in the art that methanol is required to control the SME/sulfonated fatty-acid ratios in finished commercial material (see Process section of this chapter).

A.E. Sherry's paper [55] raises the concern of the potential formation of α,β-unsaturated-γ-sultones (IUS). Sherry's paper cites a paper by D.S. Conner et al. [56] that identified 1-dodecene-1,3-sultone and 1-tetradecene-1,3-sultone as sensitizers in guinea pigs. Sherry suggests that unsaturated methyl esters used as feedstock for SMEs could also result in IUSs. Figure 19 shows the hypothetical IUS that could result from the unsaturated methylesters versus the IUS mentioned in Conner's paper. The two are substantially different from one another. No data were provided by Sherry that 1-alkyl-3-(α-sulfoalkylester)-1,3-sultone causes skin sensitization. It is also unlikely any IUS that could be formed would survive the harsh peroxide conditions used during the manufacturing of SMEs.

Finally, despite the published concerns over possible byproducts in SMEs, there are no reports that commercially available SMEs or sulfonated fatty acids have caused human skin sensitization. Studies conducted on commercial C13.6 SME/sulfonated fatty-acid blend (ALPHA-STEP MC-48) show that the product is not a skin sensitizer in either guinea pig or human testing [57].

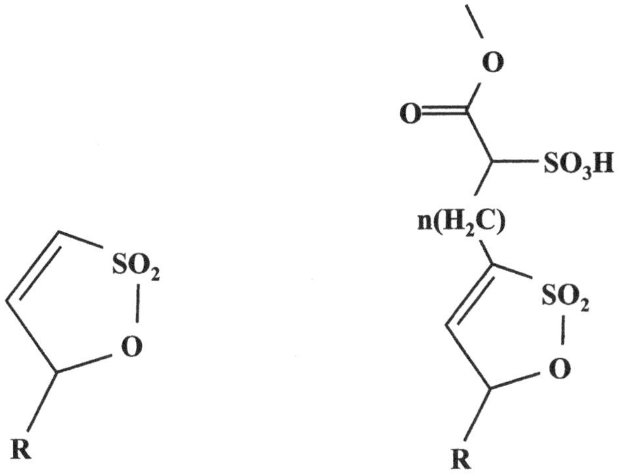

1-alkene-1,3-sultone (Conner)   α,β-unsaturated-γ-sultones (IUS)

FIG. 19   Conner's IUS versus hypothetical IUS in SMEs.

VII. FUTURE

SMEs and sulfonated fatty acids are commercially available as blends for use in household and personal cleaning products. As cleaning and personal-care formulations become more concentrated and milder, alpha-sulfo fatty derivatives will become even more attractive. Also since methyl esters are an abundant feedstock, their conversion to SMEs for use as bulk anionics is likely to grow. As quoted by Ed A. Knaggs, a former Stepan employee "As already stated, alpha sulfo fatty derivatives have exceptional and unique properties which can be varied and tailored to meet an extremely wide variety of surfactant applications. . . ." Hence, their acceptance is basically limited only by research, marketing and manufacturing dedication, imagination, innovation, commitment and support. The future potential therefore is almost limitless!

ACKNOWLEDGMENTS

The author would like to acknowledge Stepan Company for access to their data and the contributions of Marshall Nepras of Stepan Company and Teruhisa Satsuki of Lion Corporation.

REFERENCES

1. JK Weil, RG Bistline Jr, AJ Stirton. JACS 75:4859–4860, 1953.
2. T Satsuki. INFORM 3:10, 1992.
3. F Schambil, MJ Schwuger. Tenside Surf Det 27:380–385, 1990.
4. AJ Stirton, JK Weil, RG Bistline Jr. JAOCS 31:13–16, 1954.
5. W Stein, H Baumann JAOCS 52:323–329, 1975.
6. K Hovda. Proceedings of Porim International Palm Oil Congress, 1993.
7. I Yamane, Y Miyawaki. Proceedings of 1989 International Palm Oil Development Conference, Malaysia, 1989.
8. S Ahmad, Z Isail, A Rafiei, Z Zainudin, H Cheng. Proceedings of Porim International Palm Oil Congress, 1993.
9. H Yoshimura, Y Mandai, S Hashimoto. Jpn J Oil Chem Soc 41:1041, 1992.
10. BL Kapur et al. JAOCS 55:549, 1978.
11. N Nagayama, O Okumura, et al. Yukagaku 24(6):331, 1975.
12. T Ogoshi, Y Miyawaki. JAOCS 62:331, 1985.
13. K Hovda. The Challenge of Methylester Sulfonation. AOCS presentation given in Seattle, 1997.
14. F Schambil, MJ Schwuger. Tenside Surf Det 27:380–385, 1990.
15. AJ Stirton, RG Bistline Jr, JK Weil, WC Ault, EW Maurer. JAOCS 39:128–131, 1962.
16. AJ Stirton. JAOCS 39:490–496, 1962.
17. T Satsuki, K Umehara, Y Yoneyama. JAOCS 69:672–677, 1992.
18. YK Rao et al. Effect of Magnesium Counter-ion on the Surface Activity and Cleaning Efficiency of some alpha-Sulfo Methyl Ester Surfactants. CESIO presentation given in Barcelona, Spain, 1996.
19. J Ross, GD Miles. Oil Soap 18:99–102, 1941.
20. EA Knaggs, JA Yeager, L Varenyi, E Fischer. JAOCS 42:805–810, 1965.
21. JK Weil, RG Bistline Jr, AJ Stirton. JAOCS 34:100–103, 1957.
22. U.S. patent 5,616,781.
23. CZ Draves, OL Sherburne. Am Dyestuff Rep 39:771–773, 1950.
24. L Shapiro. Am Dyestuff Rep 39:38–45, P 62, 1950.
25. JC Drozd. Proceedings of 1990 World Conference of Oleochemicals, AOCS publisher, 1991:256–268.
26. NR Smith. happi March 1989:58–60.
27. A Malik, N Rockwell, YK Rao. New Horizons AOCS/CSMA Detergent Industry Conference 1995, AOCS Press, 1996:139–148.
28. HC Borghetty, CA Bergman. JAOCS 27:88–90, 1950.
29. Previously unpublished data generated by Stepan Company in 1973.
30. N Rockwell, YK Rao. Proceedings of the World Conference on Lauric Oils: Sources, Processing and Applications AOCS Press, 1994:138–146.
31. AJ Stirton, JK Weil, AA Stawitzke, S James. JAOCS 29(5):198–201, 1952.
32. G Wallace, B Sajic. Previously unpublished data generated in Stepan Company, 1994.
33. RG Bristline Jr, WR Noble, JK Weil, WM Linfield. JAOCS 49:63–69, 1972.
34. RM Anstett, EJ Schuck. JAOCS 43:576–580, 1966.

35. Previously unpublished data generated in France by Stepan Europe.
36. U.S. patent 5,637,758.
37. B Sajic, I Ryklin, B Frank. happi 34(4):94–100, 1998.
38. RL Mayhew, CF Jelinek, A Stefcik. SSCS 31(7):37, 1955.
39. Modified ASTM method, designation D-4488-91, Section A-5.
40. Previously unpublished data generated in Stepan Company.
41. U.S. patent 2,303,212.
42. German patent DE 2403895 (1974).
43. U.S. patent 3,247,121 (1996).
44. U.S. patent 5,965,508 (1999).
45. EW Maurer, JK Weil, WM Linfield. JAOCS 54:582–584, 1977.
46. M Masuda, H Odake, K Miura, K Ito, K Yamada, K Oba. JJOCS 42:905–909, 1993.
47. HA Painter. Anionic surfactants. In Handbook of Environmental Chemistry, Vol. 3, Part F, 1992:1–88.
48. RR Birch et al. Role of Anaerobic Biodegradability in the Environmental Acceptability of Detergent Materials. CESIO Conference, London, 1992.
49. JL Berna et al. Tenside Detergents 26:101–107, 1989.
50. DC McAvoy et al. Environ Toxicol Chem 1992, in press.
51. RA Rapaport et al. The Fate of Commercial LAS in the Environment. Cesio Conference, London, 1992.
52. SU Hong. Assessment of Environmental Impact and Safety of Synthetic Detergents in Korea. Korean Society of Water Pollution Research and Control, First International Symposium of the Synthetic Detergent, Seoul, 1992.
53. Von P Gode, W Guhl, J Steber. Okologische Betertung von a-Sulfofettsauremethylestern. Fat Sci Technol 89(13):548–552, 1987.
54. DW Roberts, CJ Clemett, CD Saul, A Allan, RA Hodge. Comun J Com Esp Deterg 26:27–33, 1995.
55. AE Sherry, BE Chapman, MT Creedon, JM Jordan, RL Moese. JAOCS 72:835–841, 1995.
56. DS Conner, HL Ritz, RS Ampulski, HG Kowollik, P Lim, DW Thomas, R Parkhurst. Fette Seifen Anstrichmettel 77:25–30, 1975.
57. Studies were conducted by Safepharm Laboratories Limited in the United Kingdom and Stephens & Associates Inc. in Texas, respectively.

4

Detergency Properties of Alkyldiphenyl Oxide Disulfonates

LISA QUENCER   The Dow Chemical Company, Midland, Michigan

T. J. LOUGHNEY   Samples LLC, Port Orchard, Washington

I. INTRODUCTION

Alkyldiphenyl oxide disulfonates were first identified in 1937 by Prahl [1] of E.I. du Pont de Nemours & Company. They were produced by condensation of a sulfuric acid ester of an aliphatic or cycloaliphatic alcohol with diphenyl oxide. Because of the large excess of sulfuric acid ester necessary to drive the reaction to completion, reaction yields were low. Only a few investigations followed this initial work [2,3]. In 1958 Steinhauer [4] was able to develop a commercially viable process. This made it possible to evaluate alkyldiphenyl oxide disulfonates in various applications where the basic physical properties of the chemistry were hypothesized to be of value. The original observations showed that branched alkyldiphenyl oxide disulfonates were much more soluble in electrolytes than alkylbenzene(mono)sulfonates and were in fact more soluble than other conventional surfactants of the time. The process developed by Steinhauer [4] produced alkyldiphenyl oxide with 1 to 1.3 alkyl substituents on the diphenyl oxide intermediate. Reaction of this mixture with either chlorosulfonic acid or sulfur trioxide produced the disulfonates. This reaction was carried out as a batch process using a nonreactive solvent such as a liquid polychlorinated aliphatic hydrocarbon (i.e., methylene chloride, carbon tetrachloride, perchloroethylene, or ethylene dichloride) as diluent and heat exchange agent.

The Dow Chemical Co. has been producing the alkyldiphenyl oxide sulfonates since 1958 under the trade name of Dowfax Surfactants. These products have been distributed globally for over 35 years and have capitalized on the solubility, rinsibility, and stability of this family of surfactants. Alkyldiphenyl oxide disulfonates are also produced and sold by Kao, Pilot Chemical Co., Rhodia, and Sanyo. In addition, a number of smaller suppliers are located in China, India, and France.

Traditional uses of the alkyldiphenyl oxide disulfonates include (1) emulsion polymerization to yield faster run times with less reactor waste and smaller particle size; (2) acid dyeing of nylon carpet fiber as leveling agents to promote an even distribution of dye, (3) crystal habit modification to alter crystal shape and size; and (4) cleaning formulations to provide solubilization in strongly acidic, caustic, and bleach environments.

II. STRUCTURE AND METHOD OF PREPARATION

Alkyldiphenyl oxide disulfonate surfactants are a Friedel-Crafts reaction product of an olefin and diphenyl oxide using $AlCl_3$ as a catalyst, as indicated in Figure 1. Diphenyl oxide is present in excess and is recycled. The reaction yields a mixture of monoalkyl and dialkyldiphenyl oxide. The ratio of mono- to dialkylation can be optimized depending on the end use of the products.

The next step in the process is the reaction of the alkylate with the sulfonating agent. This reaction (Fig. 1) is conducted in a solvent to dilute the reactant and to act as a diluent for the SO_3 used in the reaction. The reaction generally yields a mixture of monosulfonate and disulfonate. The level of disulfonation is determined by the end use of the product. Generally, the disulfonation level is >80%. The four primary components are shown in Figure 1. The predominant component in the commercial reaction mixture is the monoalkyldiphenyl oxide disulfonate (MADS).

The process to produce these types of materials has undergone a number of changes and modifications over the years. The Dow Chemical Company first developed a continuous process to alkylate, sulfonate, and neutralize these materials, but due to operational difficulties this process was abandoned in favor of a batch operation. The general process today (Fig. 2) consists of the reaction of an unsaturated hydrocarbon such as an alpha-olefin in the range of 6 to 16 carbons with diphenyl oxide in the presence of $AlCl_3$. The ratio of mono- to dialkylation is controlled by the ratio of olefin to diphenyl oxide. Recycled excess diphenyl oxide is purified and reused. The rate of the reaction and the yield are controlled by the amount of catalyst and temperature of the alkylation. Excessively high temperatures as well as excessive amounts of catalyst yield higher levels of di- and trialkylation. Low temperatures result in a low conversion of olefin. The ratios of concentration, catalyst, and temperature are critical in keeping the reaction products consistent throughout the production cycle. The catalyst is removed from the process stream and the crude reaction mixture is then stripped of excess diphenyl oxide. Additional purification may take place prior to the sulfonation reaction.

Sulfonation is generally carried out in a solvent. The solvent is used (1) to distribute the sulfonating agent such as SO_3, preventing localized burning and yield loss of the reaction product, and (2) to act as a heat removal medium, allowing

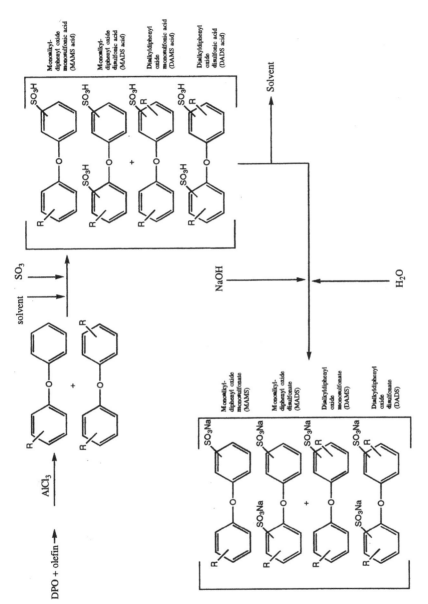

FIG. 1 Chemistry of the alkyldiphenyl oxide disulfonates. Olefin = C_{6-16} alpha olefin or tetrapropylene.

PROCESS FLOW SHEET

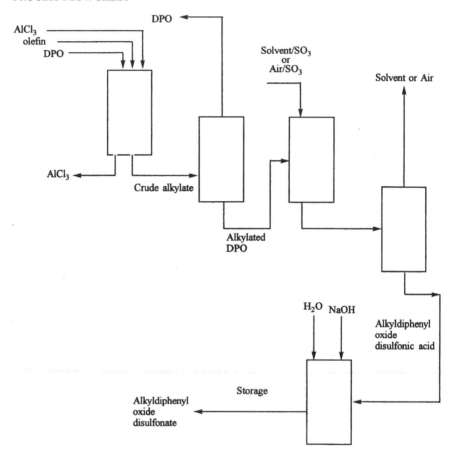

FIG. 2 Process flow sheet for the manufacture of alkyldiphenyl oxide disulfonates. DPO = diphenyl oxide.

control of the temperature of the reaction process. The commercial process routes use sulfur dioxide (Pilot Chemical Company) and methylene chloride as the preferred reaction solvents or air (Chemithon Corporation). The air sulfonation process eliminates the need for the removal and recycle of the liquid reaction solvent and is amenable to onsite generation of SO_3. Liquid solvents require the use of liquid SO_3 that is diluted into the solvent prior to addition to the sulfonation reactors. Sulfur trioxide and chlorosulfonic acid are the two most common sulfonating agents.

After sulfonation, the sulfonic acid is separated from its diluent and the anhydrous acid is then diluted with water prior to neutralization with an alkaline base such as sodium hydroxide. In the past, much of the product has been sold as the sodium salt because its viscosity is lower than that of the neat acid. This is the reverse of monosulfonated linear alkylbenzene, where the viscosity of the neat acid is lower than that of the sodium salt. The reduced viscosity of the sodium salt form is hypothesized to be due to the increased solubility of the disulfonate over that of the monosulfonate. Recent improvements in process and handling technology have resulted in high-active acid products.

The material is then packaged and sold in drums or bulk shipments as the customer may require. Storage precautions are designed to prevent the product from freezing. Upon the onset of freezing, crystals form which settle to the bottom of the container and, until thawed, will block pumps and valves. This happens at about 4°C and the surfactant solution freezes at around −1 to −3°C. Upon thawing, the disulfonated surfactant becomes homogeneous with minimal agitation.

III. DETERGENCY OF ALKYLDIPHENYL OXIDE DISULFONATES

A. Removal of Particulate Soil with Alkyldiphenyl Oxide Disulfonates

The predominant nonmechanical mechanism of removal of particulate soil from a substrate is believed to involve negative electrical potential at the Stern layers of both soil and substrate as a result of adsorption of anions from the wash bath [5]. The presence of anionic surfactants in the wash bath aids in increasing the negative potential of both the substrate and the particulate soil. With increased negative potential, the mutual repulsion between the soil and substrate is increased and the energy barrier for removal of the soil from the substrate is decreased. Along with the lowering of the energy barrier for particulate soil removal, the energy barrier for particulate soil redeposition is increased [6]. The anionic surfactants added to detergent formulations for greater removal of particulate soils are typically monosulfonated. It seems reasonable that addition of a disulfonated surfactant would lower the energy barrier even further than a typical monosulfonated surfactant.

Figure 3 illustrates the detergency performance of a linear C_{16} alkyldiphenyl oxide disulfonate compared to a typical monosulfonated surfactant, C_{12} linear alkylbenzenesulfonate. The alkyldiphenyl oxide disulfonate was observed to yield a higher detergency score than the monosulfonated surfactant over a broad concentration range. The results illustrated in Figure 3 represent a direct comparison between a mono- and disulfonated surfactant in an unformulated system at a hardness level of 150 ppm Ca^{2+}.

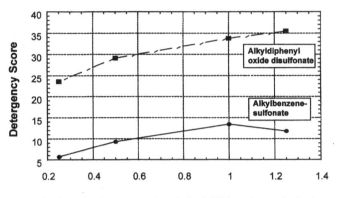

Surfactant Concentration (g/l of 35% active solution)

FIG. 3 Comparison of the detergency performance of C_{12} linear alkylbenzenesulfonate and C_{16} alkyldiphenyl oxide disulfonate. The detergency comparison was conducted at 25°C using a Tergotometer and dust/sebum-soiled swatches of cotton, cotton/polyester, and polyester. Clean swatches of each fabric type were also included to measure redeposition. The detergency score is a summation of reflectance deltas of the clean and soiled swatches before and after washing.

Investigations conducted by Rouse et al. [7] and Kumar et al. [8] have shown that monosulfonated surfactants readily precipitate in the presence of divalent counterions. In contrast, Rouse [7] found that disulfonated surfactants exhibit excellent solubility in the presence of divalent counterions such as Mg^{2+} and Ca^{2+}. Investigations by Cox et al. [9] have shown that the detergency performance of a monosulfonated surfactant decreases with increasing water hardness in the absence of components such as micelle promotion agents or builders or increased surfactant concentrations. The alkyldiphenyl oxide disulfonate was observed to provide higher detergency scores than the alkylbenzene(mono)sulfonate over a broad range of hardness (Fig. 4). Moreover, the detergency performance of the disulfonated surfactant actually increased with increasing water hardness. Both the calcium and magnesium salts of the alkyldiphenyl oxide disulfonate were observed to provide improved detergency performance over that of the sodium salt. Detergency scores of 7.5, 16.5, and 21.8 were obtained for the sodium, calcium, and magnesium salts, respectively. The detergency comparison of the different salts was conducted on cotton, cotton/polyester, and polyester fabrics.

As long as the surfactant remains soluble, this increase in detergency performance with increasing water hardness is not unexpected in view of the degree of binding of the counterion. This binding increases with an increase in polarizability and valence of the counterion. Increased counterion binding essentially

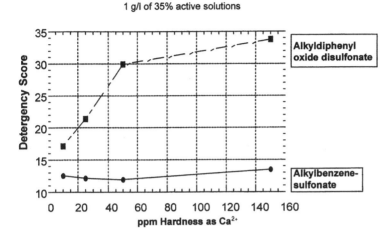

1 g/l of 35% active solutions

FIG. 4 Comparison of the detergency performance of C_{12} linear alkylbenzenesulfonate and C_{16} alkyldiphenyl oxide disulfonate at varying hardness levels. The detergency comparison was conducted at 25°C using a Tergotometer and dust/sebum-soiled swatches of cotton. cotton/polyester, and polyester. Clean swatches of each fabric type were also included to measure redeposition. The detergency score is a summation of reflectance deltas of the clean and soiled swatches before and after washing.

makes the surfactant more hydrophobic in nature. According to Rosen [10], in aqueous media for lauryl sulfate the critical micellization concentration decreases in the order $Li^+ > Na^+ > K^+ > Ca^{2+}, Mg^{2+}$.

B. Optimization of Disulfonate/Nonionic Blends for Enhanced Oily Soil Removal

While anionic surfactants are typically recognized for providing good particulate soil detergency in a formulation, nonionic surfactants are recognized for contributing good oily soil detergency. Anionic and nonionic surfactants are therefore often formulated together in order to optimize the performance of a formulation on both types of soils.

Raney and Miller [11] related the detergency performance of alcohol ethoxylate systems to their phase behavior. Raney [12] then extended that work to demonstrate the importance of the phase inversion temperature (PIT) in optimizing a blend of nonionic and anionic surfactants for oily soil removal. A similar study was undertaken to optimize a blend of commercial grade nonionic surfactant, $C_{12/13}$ 3-EO alcohol ethoxylate, and C_{16} alkyldiphenyl oxide disulfonate for the removal of cetane.

A phase inversion plot was developed for the C_{16} alkyldiphenyl oxide disulfonate/alcohol ethoxylate blends at a temperature of 35°C (Fig. 5). The y-intercept in this plot is equivalent to the optimal composition of the surfactant film. This composition provides the lowest oil-water interfacial tension at a given temperature. If the ratio of nonionic surfactant to anionic surfactant is lower than the optimum, the system is too hydrophilic in nature and tends to form an oil-in-water emulsion. Similarly, if the ratio of nonionic surfactant to anionic surfactant is higher than the optimum, a water-in-oil emulsion forms. The plot shown in Figure 5 was developed by using electrical conductivity measurements as outlined by Raney [12] to identify the transition from a high-conductivity oil-in-water emulsion to a low-conductivity water-in-oil emulsion.

As shown, a blend of 82.2% $C_{12/13}$ 3-EO alcohol ethoxylate and 17.8% C_{16}

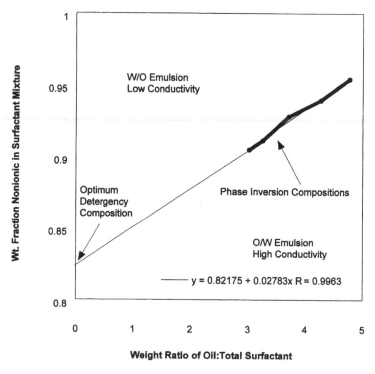

FIG. 5 Phase inversion compositions for blends of nonionic ($C_{12/13}$ 3-EO alcohol ethoxylate) and anionic (C_{16} alkyldiphenyl oxide disulfonate) surfactants at 35°C.

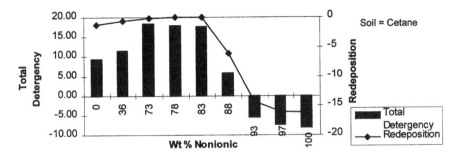

FIG. 6 Cetane removal/redeposition by blends of $C_{12/13}$ 3-EO alcohol ethoxylate and C_{16} alkyldiphenyl oxide disulfonate. The detergency evaluations were conducted at 35°C in 1% NaCl with no added hardness.

alkyldiphenyl oxide disulfonate is predicted to give the optimum detergency for removal of cetane at a temperature of 35°C. A similar plot incorporating C_{12} linear alkylbenzenesulfonate in place of the dilsulfonated surfactant predicted an optimum detergency composition of 61.4% nonionic surfactant and 38.6% anionic surfactant. These numbers indicate that a higher concentration of monosulfonated surfactant is required to invert the W/O emulsion to a O/W emulsion.

Figure 6 illustrates the detergency performance of blends of $C_{12/13}$ 3-EO alcohol ethoxylate and C_{16} alkyldiphenyl oxide disulfonate on polyester fabric soiled with cetane. The detergency of the nonionic/disulfonated anionic blends peaked at ratios close to the predicted optimum. The lowest level of redeposition of the cetane onto the clean fabric swatches was observed to occur at the predicted optimized blend. Ratios of nonionic to disulfonated anionic surfactant that yielded water in oil emulsions in Figure 5 were observed to yield the lowest detergency scores in Figure 6. Redeposition was also dramatically increased for these high ratios of nonionic surfactant. Figure 6 illustrates that oily soil can be effectively removed with a blend of nonionic and disulfonated anionic surfactant. However, the ratio at which the disulfonated anionic and nonionic surfactants are blended together should be carefully chosen.

C. Rinsability of Alkyldiphenyl Oxide Disulfonates

It is expected that standards for reduced energy and water usage proposed by the U.S. Department of Energy (DOE) will result in major changes in the design of washing machines within the United States [13,14]. These changes will require detergents that are formulated for lower water volumes and reduced washing temperatures. A high degree of surfactant solubility is required for both conditions. The dual hydrophilic headgroups of disulfonated surfactants enhance their solubility in aqueous environments and therefore should enhance their rinsability from hydrophobic surfaces. In order to test this hypothesis, the rinsability of a monosulfonated surfactant, a disulfonated surfactant, and a nonionic surfactant from a cotton substrate were compared. Cotton skeins specified in ASTM D-2281-68 [15] were allowed to completely wet in 0.0125 M solutions of C_{12} linear alkylbenzenesulfonate, C_{16} alkyldiphenyl oxide disulfonate, nonylphenol ethoxylate, and sodium xylenesulfonate. The sodium xylene sulfonate was included as a control since, because it lacks a large hydrophobe, it was not expected to adhere to the cotton substrate. After the skeins were wet in the 0.0125 M surfactant solutions, they were packed in chromatography columns. The void space in the chromatography columns was filled with the equilibrium surfactant solution in which the skeins were prewet. Distilled water was then pumped through the columns at a rate of 5 mL/min. The effluent from the columns was collected and analyzed using a UV-VIS spectrophotometer. The quantity (in mols) of each surfactant rinsed from the cotton substrate as a function of time is shown in Figure 7.

As predicted, sodium xylenesulfonate was not substantive to the cotton substrate. The nonionic nonylphenol ethoxylate was observed to be the most substantive to the cotton fiber of the surfactants evaluated. The disulfonated surfactant rinsed more readily from the cotton fiber than the monosulfonated sur-

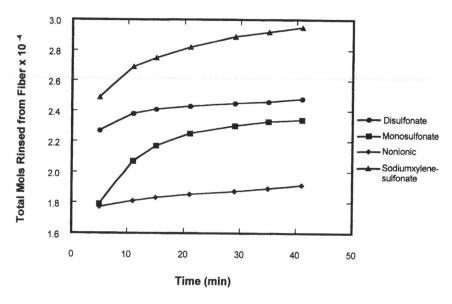

FIG. 7 Mols of surfactants rinsed from cotton skeins as a function of time. Disulfonate = C_{16} alkyldiphenyl oxide disulfonate; monosulfonate = C_{12} linear alkylbenzenesulfonate; nonionic = nonyl phenol ethoxylate (EO = 9). All rinsability studies were conducted at ambient temperature.

factant. Since all of the surfactants were compared on an equal molar basis, in actual use the rinsibility of the disulfonated surfactant would be even more favorable because of the large difference in molecular weight between the disulfonated and monosulfonated molecules (643 g/mole vs. 348 g/mole). Typically the surfactants are used on a weight basis rather than a mole basis. The detergency performance of the two surfactants on an equal weight basis has already been illustrated in Figure 3.

D. Foaming

The foam profiles of surfactants used in detergent formulations will become even more important as the anticipated changes in washing machine design occur [16]. One of the anticipated changes in machine design is a switch from the typical vertical-axis machines used throughout the United States today to horizontal-axis machines similar to those currently used throughout Europe. The anticipated horizontal axis design coupled with an increased surfactant concentration due to lower water volumes will require formulations with low foam profiles. The foaming of a formulation may be reduced by either adding an antifoam agent to the formulation or by formulating with components that do not contribute to foaming.

The alkyldiphenyl oxide disulfonates are considered moderate foamers. While foam can be generated upon agitation of the surfactant solution, the foam profile tends to be lower than that of many other common monosulfonated surfactants. Figure 8 illustrates modified Ross-Miles [17] foam profiles of the C_{12} linear alkylbenzenesulfonate and of C_{16} alkyldiphenyl oxide disulfonate. The disulfonated surfactant was observed to yield a lower foam profile over the entire range of conditions evaluated. The foam profile of the disulfonated surfactant, C_{16} alkyldiphenyl oxide disulfonate, varies with the solution in which it is solubilized. For example, the initial foam height of the disulfonated surfactant in the presence of 5% sodium hypochlorite is approximately one-half of that observed in deionized water (Table 1).

E. Multifunctionality

The high tolerance of C_{16} alkyldiphenyl oxide disulfonate to divalent cations has been illustrated in a previous section. Because of their increased solubility, the shorter-chained alkyldiphenyl oxide disulfonates, C_6 and C_{10}, show an even greater tolerance toward divalent cations. The ability of these surfactants to withstand hard-water conditions can result in a reduction of additives such as builders and chelates to a formulation. The alkyldiphenyl oxide disulfonates can also exhibit multifunctionality in a formulation by contributing to detergency performance and at the same time enhancing the solubilization of other formulation ingredients. Figure 9 illustrates the solubilization of nonylphenol ethoxylate

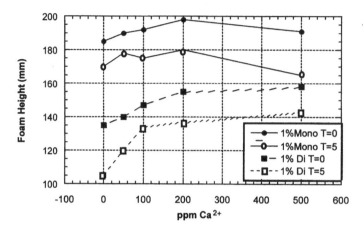

FIG. 8 Modified Ross-Miles foam heights for C_{12} linear alkylbenzenesulfonate and C_{16} alkyldiphenyl oxide disulfonate at varying levels of Ca^{2+}. Mono = C_{12} linear alkylbenzenesulfonate; Di = C_{16} alkyldiphenyl oxide disulfonate. T = 0 represents the initial foam height. T = 5 represents the foam height after 5 min.

TABLE 1 Ross Miles Foam Height Measurements for 1% Active C_{16} Alkyldiphenyl Oxide Disulfonate in Various Solutions at 25°C

Solution	Initial foam height	Foam height at 5 min
Deionized water	120 mm	40 mm
5% sodium hypochlorite	60 mm	0 mm
20% sodium hydroxide	63 mm	56 mm
20% phosphoric acid	120 mm	45 mm

by C_6 and C_{16} alkyldiphenyl oxide disulfonates in an alkaline environment. The longer-chain C_{16} alkyldiphenyl oxide disulfonate provides detergency performance to the formulation and at the same time serves to solubilize the nonionic surfactant. Although the shorter-chain C_6 alkyldiphenyl oxide disulfonate is not as effective in detergency as the longer-chain C_{16} surfactant, it is the most effective coupler.

F. Bleach Stability

Sodium hypochlorite, bleach, has been a staple ingredient of cleaning products for well over a hundred years. The extensive use of bleach is due to its ability to decolorize stains, oxidize some soils thereby promoting their removal, and provide antibacterial activity against a wide variety of microorganisms. Alone,

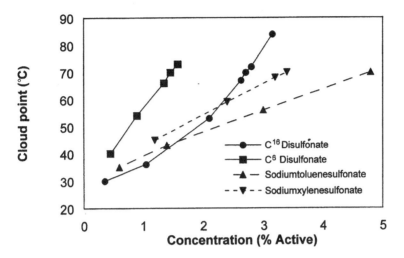

FIG. 9 Cloud point modification of nonyl phenol ethoxylate (EO = 9) in 5% NaOH with C_6 and C_{16} alkyldiphenyl oxide disulfonates and common hydrotropes.

bleach lacks good detergency and wetting properties. To enhance its performance, it is therefore often formulated with other ingredients, such as surfactants. Addition of selected surfactants can increase the viscosity of the sodium hypochlorite solutions which, in turn, may improve their cleaning and disinfecting properties as a result of longer contact times with the substrate, particularly on vertical surfaces. To mask the distinctive chlorine odor, fragrances are often added to solutions of sodium hypochlorite for both household and institutional use. Surfactants are useful in this context also since they promote the solubilization of fragrance into bleach solutions. Even though the addition of surfactants and fragrances result in performance and esthetic improvements, both materials must be chosen with care to avoid interaction with sodium hypochlorite which results in loss of both the additive as well as of the sodium hypochlorite. The stability of the sodium hypochlorite may also be affected by additional factors (i.e., storage temperature, metal content, pH, etc.) which vary from formulation to formulation.

The stability of sodium hypochlorite in the presence of different surfactants was evaluated by preparing solutions in distilled water containing 4.5% sodium hypochlorite, 1.0% active surfactant, and 0.3% sodium hydroxide. The concentration of store-brand sodium hypochlorite bleach was determined as percent available chlorine according to an ASTM procedure [18]. The remaining formulation ingredients were then added directly to the sodium hypochlorite solution.

Subsequent measurements were all recorded as percent available chlorine using the cited ASTM titration method. All samples were stored in the dark at 37°C.

As seen in Table 2, sodium hypochlorite bleach maintains excellent stability in the presence of the alkyldiphenyl oxide disulfonates, even after 319 days. Only 2-ethylhexylsulfate exhibited similar inertness to sodium hypochlorite. The bleach solution fared moderately well with a number of surfactants to the 137-day point, but were not tested at 319 days due to the level of precipitate that developed. Two of the surfactants, phosphate esters and alkyl polysaccharide, exhibited very poor stability.

G. Detergency Properties of the Components of Alkyldiphenyl Oxide Disulfonate Reaction Mixtures

The components (MAMS, MADS, DADS, DAMS; Fig. 1) that make up the commercial blend of alkyldiphenyl oxide disulfonates were evaluated individually in both built and unbuilt formulations to determine the effect of equivalent weight on detergency performance (Fig. 10) Measurements were obtained on dust/sebum and ground-in-clay soils on cotton, cotton/polyester, and polyester fabric.

The C_{16} monoalkyldiphenyl oxide disulfonate component yielded the optimum detergency in the unbuilt formulation. In built formulations, the C_{10} dialkyldiphenyl oxide disulfonate component showed optimum detergency. In

TABLE 2 Percent of Original Available Chlorine Upon the Aging of Sodium Hypochlorite Solutions in the Presence of Various Surfactants

Surfactant	137 Days	319 Days
None	92.6	84.1 clear
C_{12} alkyldiphenyl oxide disulfonate	90.8	80.6 clear
C_6 alkyldiphenyl oxide disulfaonte	88.0	76.4 clear
C_{16} alkyldiphenyl oxide disulfonate	87.0	75.8 clear
2-Ethylhexyl sulfate	86.2	75.7 clear
C_{10} alkyldiphenyl oxide disulfonate	87.1	74.3 clear
Nonylphenolethoxylate, EO = 9	85.7	73.6 hazy
Alkylnapthalenesulfonate	71.0	28.8 clear
Sodium lauryl sulfate	92.3	ppt
C_{12} linear alkylbenzenesulfonate	76.4	ppt
Lauryl ether sulfate	54.4	ppt
C_9–C_{11} alcohol ethoxylate, EO = 8	74.4	ppt
C_{14}–C_{16} alpha olefin sulfonate	57.6	ppt
Phosphate ester	81.8	0.0 clear
C_9–C_{11} Alkyl polysaccharide	0.0	0.0 clear

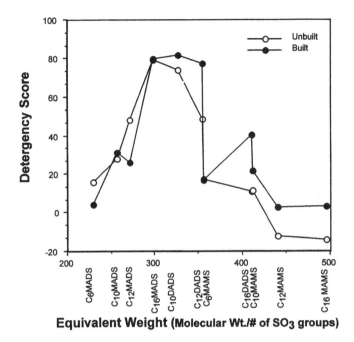

FIG. 10 Detergency performance of the individual components of commercial alkyldiphenyl oxide disulfonates as a function of equivalent weight. MADS = monoalkyldiphenyl oxide disulfonate; MAMS = monoalkyldiphenyl oxide monosulfonate; DADS = dialkyldiphenyl oxide disulfonate.

both formulation types, a disulfonated component provided the best detergency performance. A correlation between equivalent weight and detergency performance may be observed for both formulations. The shorter-chain monoalkyldiphenyl oxide disulfonates with low equivalent weights are too water soluble to provide good detergency performance. The higher equivalent-weight monoalkyldiphenyl oxide monosulfonates are too hydrophobic and yield a negative total score indicating that soil was deposited onto the clean fabric swatches. Because of their low solubility in water, the dialkyldiphenyl oxide monosulfonates were not tested.

Investigations undertaken by Rosen et al. [19, 20] have shown the dialkyldiphenyl oxide disulfonates capable of enhancing the wetting properties of water-insoluble surfactants and to provide properties such as low critical micellization concentrations and low C_{20} values [5]. The detergency properties of the dialkyldiphenyl oxide disulfonates have not yet been thoroughly investigated.

IV. ENVIRONMENTAL/SAFETY

A. Skin Irritation

Commercial-grade C_{16} alkyldiphenyl oxide disulfonate was compared to commercial grade C_{12} linear alkylbenzenesulfonate sodium salt and a commercial grade sodium C_{12-14} 2.8 to 3.0 EO alcohol ether sulfate in a study which evaluated primary dermal irritation on New Zealand white rabbits. Aliquots of 0.5 mL of 10% active surfactant were applied for 4 hours to the intact skin on the backs of groups of six New Zealand white rabbits.

With C_{16} alkyldiphenyl oxide disulfonate, no signs of dermal irritation were observed up to 3 days on the test site in any of the six rabbits that were dosed. With the alcohol ether sulfate, two of the six rabbits showed slight erythema when the test site was unwrapped approximately 30 min after dosing. The erythema was resolved at the 48-hour observation. This test, too, was terminated after 3 days. With C_{12} linear alkylbenzenesulfonate sodium salt, three of the six rabbits had a slight erythema and/or slight edema when the test site was unwrapped approximately 30 min after dosing. One of the three affected rabbits continued to show slight erythema and/or edema until test day 7. In addition, there was also scaling of the test site of this rabbit on test day 7, when the test was terminated. There was no effect on the body weights of the rabbits for any of the three test materials.

B. Degradation of C_{16} Alkyldiphenyl Oxide Disulfonates in Activated Sludge

The degradation of commercial alkyldiphenyl oxide disulfonates was evaluated in the semicontinuous activated sludge (SCAS) confirmatory test [21]. For both the C_{10} and C_{16} alkyldiphenyl oxide disulfonates, > 90% reduction in Methylene Blue reactive substances occurred allowing classification of these products as biodegradable under the conditions of the SCAS test. By contrast, the commercial C_{12} branched and C_6 alkyldiphenyl oxide disulfonates did not achieve a > 90% reduction in Methylene Blue reactive substances and are not considered biodegradable under this procedure.

Since the SCAS test does not provide information on the degradation pathway of the alkyldiphenyl oxide disulfonates, a study was undertaken to observe the degradation of these molecules. Because of the analytical complexities associated with the commercial mixture of alkyldiphenyl oxide disulfonates, degradation studies initially focused on a single isomer, the C_{16} monoalkyldiphenyl oxide disulfonate, shown in Figure 11. The choice of labeling the disubstituted ring shown in Figure 11 is supported by the expected biodegradation pathway which, in turn, is based on the known biodegradation pathway of linear alkylbenzenesulfonate [22].

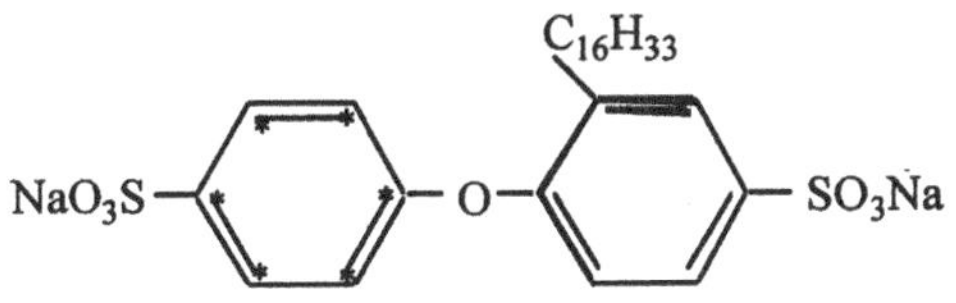

FIG. 11 Radiolabeled monoalkyldiphenyl oxide disulfonate.* Carbon 14.

Biodegradation is expected to be initiated on the terminal carbon of the linear alkyl group (omega oxidation). Degradation of the linear chain should proceed with consecutive losses of two carbon fragments (beta oxidation) to a carboxylic acid. Subsequent steps in the pathway are more speculative. One possible pathway would include opening of the trisubstituted ring followed by the degradation of the disubstituted ring. Thus, labeling of the disubstituted ring would permit a study of the fate of the C_{16} alkyldiphenyl oxide disulfonate to the point when both aromatic rings are degraded.

A fill-and-draw design modeled after the Soap and Detergent Association method [21] was used to study the degradation of the C_{16} alkyldiphenyl oxide disulfonate under both acclimated and nonacclimated conditions. Municipal activated sludge (2500 mg/L suspended solids) was acclimated to 20 ppm C_{16} alkyldiphenyl oxide disulfonate. Activated sludge not exposed to the disulfonated surfactant was maintained in a parallel system. Following acclimation, radiolabeled disulfonated surfactant was introduced into each system. Primary degradation (degradation of the parent compound) is shown in Figure 12. Even though rapid degradation occurred in the acclimated activated sludge, < 2% radiolabeled CO_2 was collected. Figure 13 illustrates the major degradation product as determined by analysis of the effluent. This degradation product supports the hypothesized degradation pathway.

The biodegradability of C_{16} alkyldiphenyl oxide disulfonate was evaluated separately under aerobic conditions using the SCAS test [21]. The surfactant passed these tests with > 90% reduction of methylene blue active substance occurring following 23 hours of aeration. In addition the surfactant showed 90% reduction of Methylene Blue active substance in 19 days in the OECD Screening and Confirmatory Test [23]. More recently, the biodegradation of the surfactant was evaluated in the OECD Zahn-Wellens/EMPA test [24]. The surfactant was shown to be inherently biodegradable according to the test guidelines since ~55% degradation was observed within 28 days.

C. Degradation in Surface and Subsurface Soils

Standard microcosms were prepared with either aquifer solids (subsurface sandy loam) or a sandy loam surface soil and dosed with radiolabeled C_{16}

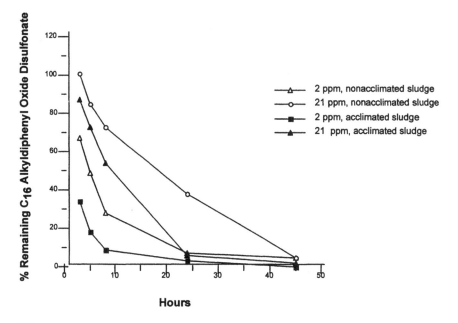

FIG. 12 Primary degradation of C_{16} monoalkyldiphenyl oxide disulfonate.

$$HOOC - CH_2 - CH - CH_3$$

NaO₃S ——⟨ ⟩—— O ——⟨ ⟩—— SO₃Na

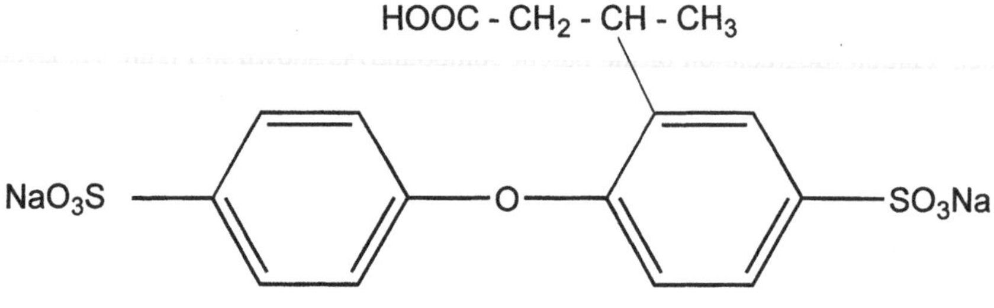

FIG. 13 Major degradation product of C_{16} monoalkyldiphenyl oxide disulfonate.

alkyldiphenyl oxide disulfonate. As shown in Table 3, significant mineralization was achieved in the sandy loam surface soil. The major degradation intermediate detected was similar to the product identified in the activated sludge study (Fig. 13).

It can be concluded that the primary biodegradation of the C_{16} alkyldiphenyl oxide disulfonate proceeds in a variety of aerobic environments (activated sludge, surface, and subsurface soils). The degradation pathway is consistent with that of the linear alkylbenzene(mono)sulfonate.

D. Aquatic Toxicity of the Major Degradation Intermediate

The acute aquatic toxicity of the major degradation intermediate (Fig. 13) was determined under static conditions on both rainbow trout and daphnid with replicate groups of 10 organisms. The results of this study are summarized in Table 4. As seen from these results, the control effluent had no effect on the rainbow trout or the daphnid. The biodegradation products from the activated sludge treated with C_{16} alkyldiphenyl oxide disulfonate (20 ppm initial concentration) had no effect on the rainbow trout or daphnid. The fact that the disulfonated surfactant, when added directly to the effluent, is toxic to fish is consistent with other sulfonated surfactants [25,26].

TABLE 3 Degradation in Soil

Soil	ppm MADS	95% Primary degradation	Mineralization* after 85 days
Subsurface sandy soil	1	10 days	<1%
Subsurface sandy soil	20	30 days	<1%
Sandy loam surface soil	1	4 days	29%
Sandy loam surface soil	20	4 days	12%

*Conversion of a compound to carbon dioxide, water and/or inorganic compounds, reported as % $^{14}CO_2$.

TABLE 4 Aquatic Toxicity of the Major Degradation Intermediate

	Species	Time (hours)	LC_{50}	EC_{50}
Blank	Trout	96	>100%*	>100%
	Daphnid	48	>100%	>100%
Intermediate	Trout	96	>100%	>100%
	Daphnid	48	>100%	>100%
C_{16} alkyldiphenyl oxide disulfonate	Trout	96	0.7 mg/L	0.7 mg/L
	Daphnid	48	14.1 mg/L	13.5 mg/L

*No toxicity.
Samples of activated sludge effluent were prepared in semi-continuous activated sludge (SCAS) units: Blank = effluent from the SCAS unit; Intermediate = feed to the SCAS unit containing 20 ppm C_{16} alkyldiphenyl oxide disulfonate (the effluent was confirmed to contain the previously identified major intermediate); C_{16} alkyldiphenyl oxide disulfonate = effluent from the SCAS unit which was amended with surfactant at known concentrations.

ACKNOWLEDGMENTS

The authors would like to thank Stanley Gonsior, Sally Kokke-Hall, Pushpa Inbasekaran, Dave Wallick, and Beth Haines of the Dow Chemical Company, and Chris Means of Kelly Services, who have provided data for inclusion in this chapter.

REFERENCES

1. MA Prahl. U.S. patent 2,081,876 to E.I. Dupont de Nemours & Company (1937).
2. GH Coleman, RP Perkins. U.S. patent 2,170,809 to Dow Chemical Company (1939).
3. CF Prutton. U.S. patent 2,555,370 to Lubrizol Corporation (1951).
4. A Steinhauer. U.S. patent 2,854,477 to Dow Chemical Company (1958).
5. MJ Rosen. In: Surfactants and Interfacial Phenomena, 2nd ed. New York: Wiley Interscience, 1989:143, 368–370.
6. W Kling, H Lange. J Am Oil Chem Soc 37(1):30, 1960.
7. JD Rouse, DA Sabatini. Environ Sci Technol 27(10):2072, 1993.
8. R Kumar, SGT Bhat. J Am Oil Chem Soc 64(4):556, 1987.
9. MF Cox, KL Matheson. J Am Oil Chem Soc 62(9): 1396, 1985.
10. MJ Rosen. In: Surfactants and Interfacial Phenomena, 2nd ed. New York: Wiley Interscience, 1989:134–135.
11. KH Raney, CA Miller. J Colloid Interface Sci 119: 539, 1987.
12. KH Raney. J Am Oil Chem Soc 68(7):525, 1991.
13. Detergents and Laundry Additives in High-Efficiency Washers. New York: Soap and Detergent Association, 1996.
14. HC Carson, T Branna. HAPPI 30(3): 72, 1993.
15. ASTM D2281-68 (reapproved 1986). Annual Book of ASTM Standards, Vol. 15.04.
16. KH Raney. In: MS Showell, ed. Powdered Detergents, Surfactant Science Series, Vol. 71. New York: Marcel Dekker, 1998:277.
17. ASTM D1173-53 (reapproved 1986). Annual Book of ASTM Standards, Vol. 15.04.
18. ASTM D 2022-87. Annual Book of ASTM Standards, Standard Methods of Sampling and Chemical Analysis of Chlorine-Containing Bleaches.
19. MJ Rosen. Chemtech March: 30, 1993.
20. MJ Rosen, ZH Zhu. J Am Oil Chem Soc 70(1):65, 1993.
21. Subcommittee on Biodegradation Test Methods of the Soap and Detergent Association. J Am Oil Chem Soc 42:986, 1965.
22. RD Swisher. In: Surfactant Biodegradation, Surfactant Science Series, Vol. 18. New York: Marcel Dekker, 1987:574–621.
23. OECD Screening and Confirmatory Tests. Organisation for Economic Co-operation and Development, 1979.
24. 302B Zahn Wellens/EMPA Test. Organisation for Economic Co-operation and De-

velopment (OECD) Guidelines for Testing of Chemicals. Adopted by the Council 17 July 1992.

25. HA Painter. In: NT de Oude, ed. Detergents, Vol. 3, Part F: Anthropogenic Compounds. New York: Springer Verlag, 1992:36.

26. LA Baker, ed. Environmental Chemistry of Lakes and Reservoirs, Advances in Chemistry Series 237. Washington: American Chemical Society, 1994: 542.

5
Methyl Ester Ethoxylates

MICHAEL F. COX and UPALI WEERASOORIYA
CONDEA Vista Company, Austin, Texas

I. INTRODUCTION

Over 2 million metric tons of ethylene oxide-derived surfactants are consumed worldwide each year [1]. They are produced by more than 150 different ethoxylators located in almost every developed country in the world.

The high consumption figure of ethylene oxide derived surfactants is due to several factors: the surfactants are relatively easy to manufacture, they are relatively inexpensive, and they can be derived from a variety of hydrophobic feedstocks, including oleochemical- and petrochemical-based alcohols, and petrochemical-based alkylphenols and amines. Because their hydrophobe and hydrophile chain lengths can be varied significantly, they fit a wide range of applications.

Historically, ethylene oxide–derived surfactants have been produced from hydrophobic feedstocks which contain an "active," or labile, hydrogen, a hydrogen that is connected to a heteroatom such as oxygen or nitrogen. For alcohols, this active hydrogen atom is readily removed with base to form a reactive anion which then reacts quickly and efficiently with ethylene oxide to form the ethoxylate. For amines, the active hydrogen atom is easily rearranged from the nitrogen to the oxygen during reaction with ethylene oxide to form the ethoxylate.

In 1989, the concept of ethoxylating methyl esters, which do not carry a labile hydrogen, was introduced by Hoechst [2] and Henkel [3]. Hoechst demonstrated that ethoxylation of esters was chemically feasible using catalysts based on alkali and alkaline earth metals (e.g., sodium hydroxide, sodium methoxide, barium hydroxide, etc.). Henkel demonstrated the feasibility of using calcined hydrotalcite (aluminum-magnesium hydroxycarbonates) for the reaction. Reactivities and conversions with these catalysts, however, were found to be too low for commercial application.

In 1990, Vista Chemical Company (now CONDEA Vista Company) developed a commercially viable process based on more complex alkoxylation catalysts (activated calcium and aluminum alkoxides) that effectively and efficiently achieved the ethoxylation of esters [4]. Soon after, Lion demonstrated that magnesium oxide–based catalysts also worked well [5]. These discoveries opened the door to ester alkoxylate development and have led to a flurry of research directed at understanding and utilizing these materials, most notably methyl ester ethoxylates [6–19].

Henkel has also demonstrated that acceptable reactivities and conversions can be achieved with hydrotalcite when cocatalysts such as ethylene glycol, fatty acids, standard alkali catalysts, etc., are used [20,21].

It is important to note that methyl ester ethoxylates have been produced, primarily for the textile industry, for several years. They have been manufactured by condensing fatty acids with monomethyl-capped polyethylene glycol. This reaction, however, is more complex and more costly than direct ethoxylation [22].

Methyl ester ethoxylates are similar in structure to conventional alcohol ethoxylates, but the structural differences that do exist have an important impact on their performance. As shown in Figure 1, methyl ester ethoxylates contain an ester linkage at the hydrophobe-hydrophile boundary of the molecule in place of the ether linkage in alcohol ethoxylates. This ester linkage sterically constrains the molecule, which reduces the tendency of the surfactant to micellize and leads to a higher critical micelle concentration (CMC). Methyl ester ethoxylates also carry a terminal methoxy group in place of a terminal hydroxyl group. This re-

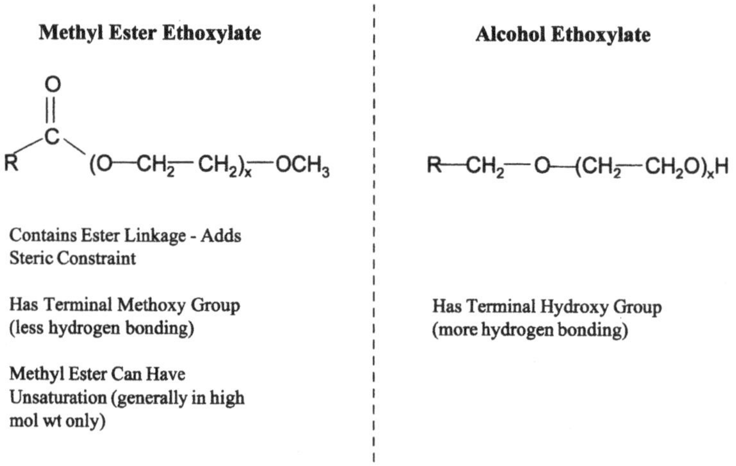

Methyl Ester Ethoxylate

$$R-\overset{\overset{\displaystyle O}{\|}}{C}-(O-CH_2-CH_2)_x-OCH_3$$

Contains Ester Linkage - Adds
Steric Constraint

Has Terminal Methoxy Group
(less hydrogen bonding)

Methyl Ester Can Have
Unsaturation (generally in high
mol wt only)

Alcohol Ethoxylate

$$R-CH_2-O-(CH_2-CH_2O)_xH$$

Has Terminal Hydroxy Group
(more hydrogen bonding)

FIG. 1   Methyl ester ethoxylates vs. alcohol ethoxylates: differences in molecular structure.

duces the hydrogen bonding of the surfactant, which in turn reduces water solubility and the tendency to form aqueous gels.

Also, unlike most alcohols, methyl esters can have a significant level of unsaturation. The degree of unsaturation varies depending on the oleochemical source of the feedstock, and the carbon chain length. In general, the higher the carbon chain length, the higher the degree of unsaturation. The impact of unsaturation on performance is addressed later in this chapter.

Another important characteristic of methyl ester ethoxylate structure is the distribution of the ethoxymers (the relative concentration of unethoxylated feedstock, of 1-mol ethoxymer, of 2-mol ethoxymer, etc.). As discussed later in this chapter, the ethoxymer distribution of methyl ester ethoxylates, like that of their alcohol ethoxylate counterparts, can vary depending on the catalyst used to prepare them.

Large quantities of methyl esters are currently produced from oleochemical sources (see Fig. 2). Their major use is as intermediates in the production of fatty alcohols. They can also be produced through esterification of fatty acids with methanol. It is safe to assume that commercial quantities of methyl esters could be made readily available as ethoxylation feedstocks for methyl ester ethoxylates.

This chapter first examines the ethoxylation of esters and the composition of methyl ester ethoxylates. Important aspects related to the formulation of liquid detergents with methyl ester ethoxylates are then discussed, followed by a re-

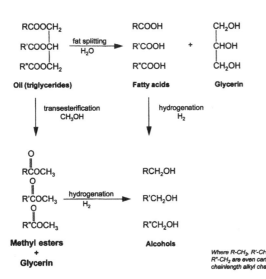

FIG. 2 Major oleochemical routes for making alcohols.

view of the performance properties of methyl ester ethoxylates, and a brief discussion of the use of methyl ester ethoxylates in household and industrial detergents. The ethoxylation of other types of esters, the propoxylation of esters, and the impact of unsaturation on performance are also briefly discussed.

II. ETHOXYLATION OF ESTERS

Ethoxylation of methyl esters using conventional hydroxide catalysts (NaOH, KOH, etc,) does not proceed efficiently because of the absence of an active hydrogen. As shown in the chromatogram of the resulting ethoxylate (Fig. 3A), conversion is poor, and the product contains a broad distribution of ethoxymers as well as significant concentration of unethoxylated methyl ester.

In contrast, ethoxylation of methyl esters with a more complex catalyst system (activated calcium and aluminum alkoxide [23]) is significantly more successful in achieving efficient ethoxylation, reducing the level of unethoxylated methyl ester, and yielding a more "peaked" distribution of ethoxymers (see Fig. 3B).

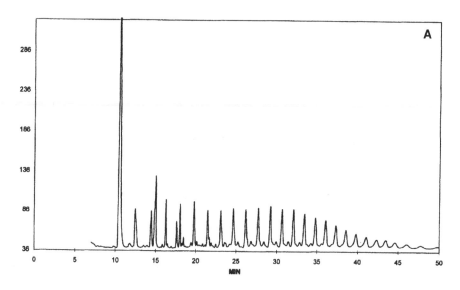

FIG. 3 A. Ethylene oxide distribution (via supercritical liquid chromatography) of C_{14} methyl ester ethoxylate obtained with conventional (NaOH) catalyst. B. Ethylene oxide distribution (via supercritical liquid chromatography) of C_{14} methyl ester ethoxylate obtained with Ca/Al-alkoxide catalyst. (From Ref. 23.) C. Ethylene oxide distribution (via supercritical liquid chromatography) of C_{14} alcohol ethoxylate obtained with conventional (NaOH) catalyst. D. Ethylene oxide distribution (via supercritical liquid chromatography) of C_{14} alcohol ethoxylate obtained with Ca/Al-alkoxide catalyst. (From Ref. 23.)

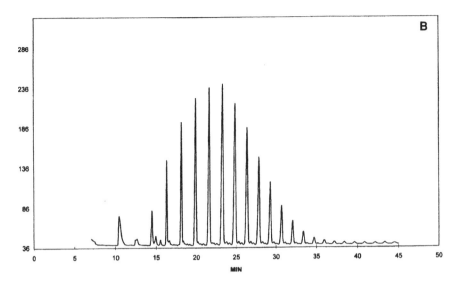

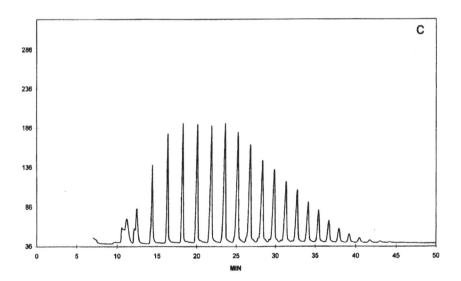

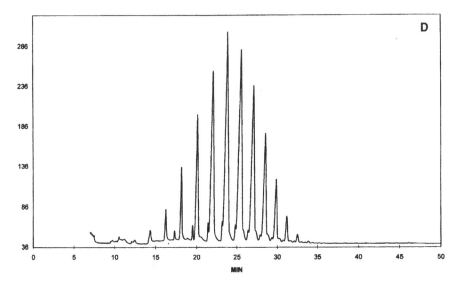

FIG. 3 Continued

These proprietary catalysts are more expensive because they must be manufactured rather than simply purchased. Certain catalysts must also be removed
from the finished ethoxylate via filtration in order to obtain a clear, transparent
ethoxylate. On the other hand, complex catalysts can be significantly more reactive than hydroxide catalysts so that the ethoxylation reaction proceeds in
less time, with less catalyst, and at lower temperatures. Ethoxylation with
complex catalysts is accomplished via conventional ethoxylation equipment
and procedures.

The mechanism of the ethoxylation of esters with these complex catalysts is
not well understood. It is thought to involve a transesterification which effectively inserts ethylene oxide into the ester linkage between the carbonyl carbon
and the methoxy oxygen [9,15,18]. This mechanism is illustrated in Figure 4. As
shown, the active catalyst (calcium and aluminum alkoxyethoxylate) first reacts
with ethylene oxide to form the ethoxylated version of the metal alkoxyethoxylate. This molecule then transesterifies with methyl ester to form the alkyl ester
ethoxylate and a metal-coordinated methoxide ion. Addition of more ethylene
oxide (step 2) produces progressively more highly ethoxylated versions of the
metal-coordinated methoxide ions, which then transesterify with the ester (step
3) to form methyl ester ethoxylate, the alkyl ester ethoxylate, and the metal-coordinated methoxide. Steps 2 and 3 occur continuously with the addition of more
ethylene oxide until excess methyl ester is consumed. This results in a distribu-

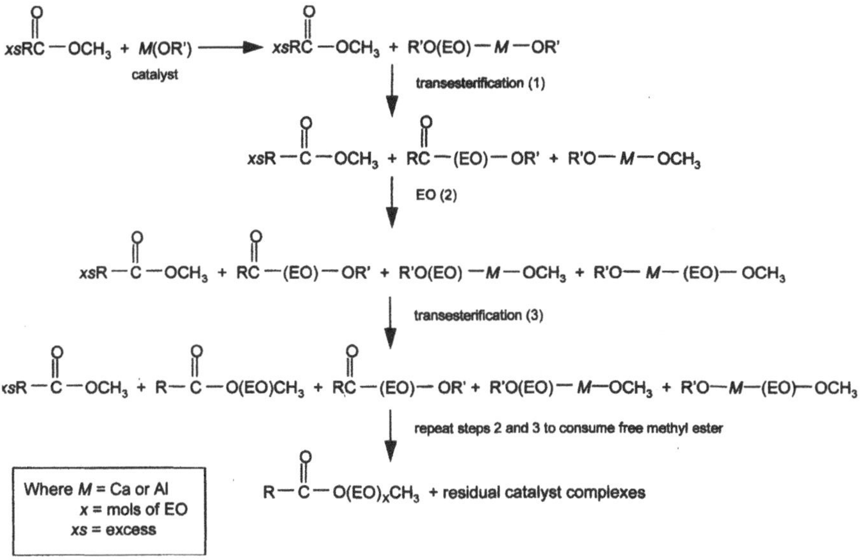

$$xsR\overset{O}{\underset{||}{C}}-OCH_3 + M(OR') \xrightarrow{\text{catalyst}} xsR\overset{O}{\underset{||}{C}}-OCH_3 + R'O(EO)-M-OR'$$

transesterification (1)

$$xsR-\overset{O}{\underset{||}{C}}-OCH_3 + RC-\overset{O}{\underset{||}{}}(EO)-OR' + R'O-M-OCH_3$$

EO (2)

$$xsR-\overset{O}{\underset{||}{C}}-OCH_3 + RC-\overset{O}{\underset{||}{}}(EO)-OR' + R'O(EO)-M-OCH_3 + R'O-M-(EO)-OCH_3$$

transesterification (3)

$$xsR-\overset{O}{\underset{||}{C}}-OCH_3 + R-\overset{O}{\underset{||}{C}}-O(EO)CH_3 + RC-\overset{O}{\underset{||}{}}(EO)-OR' + R'O(EO)-M-OCH_3 + R'O-M-(EO)-OCH_3$$

repeat steps 2 and 3 to consume free methyl ester

Where M = Ca or Al
x = mols of EO
xs = excess

$$R-\overset{O}{\underset{||}{C}}-O(EO)_xCH_3 + \text{residual catalyst complexes}$$

FIG. 4 Proposed mechanism for the alkoxylation of methyl ester ethoxylates.

tion of methyl ester ethoxylates containing a small concentration of residual catalyst complexes.

Few studies have yet been published which investigate this ethoxylation mechanism in detail. Hama, however, has investigated the impact of catalyst structure on catalyst activity and product composition for magnesium oxide–based catalysts [12].

Variations in methyl ester purity or composition (unsaturation, carbon chain length distribution, etc.) do not appear to influence the ethoxylation reaction or overall ethoxylate quality. The purity of the methyl ester, however, does appear to impact color. A yellow-tinted methyl ester logically yields a yellow-tinted ethoxylate.

III. COMPOSITION OF METHYL ESTER ETHOXYLATES

Ethoxylates can be prepared from any methyl ester. Normally, methyl esters are derived from oleochemical sources, and the carbon chain length distribution and the level of unsaturation can vary significantly depending on the specific feedstock used and the quality of the distillation process employed.

The distribution of methyl ester ethoxymers (the relative concentrations of unethoxylated feedstock, of 1-mole ethoxylate, of 2-mole ethoxylate, etc.) depends on the catalyst and conditions used. A calcium/aluminum alkoxide catalyst

[16] yields ethoxymer distributions which fall between what is typically considered conventional or "broad" (the ethylene oxide chain distribution of alcohol ethoxylates produced with conventional alkali hydroxide catalysts) and what is considered "peaked" or "narrow-range" (the ethylene oxide chain distribution of alcohol ethoxylates produced with calcium and aluminum alkoxide catalysts). Figures 3A–D illustrate ethoxymer distributions in C_{14} linear, primary alcohol ethoxylate, and C_{14} methyl ester ethoxylate produced with a conventional sodium hydroxide catalyst and with a calcium/aluminum alkoxide catalyst. As shown, sodium hydroxide effectively ethoxylates the alcohol but is relatively less effective in the ethoxylation of the methyl ester. In contrast, the complex catalyst can effectively ethoxylate either feedstock, although the degree of peaking achieved with the methyl ester is less than that achieved with fatty alcohol.

Hama has demonstrated that magnesium oxide catalysts can be used to produce both broad and peaked methyl ester ethoxylates [13]. The impact of peaking the ethoxymer distribution of methyl ester ethoxylates has also been studied by Hama [13]. He demonstrated that increased peaking reduces the tendency of the ethoxylate to form gels, reduces inverse cloud point (which is noteworthy since the opposite trend is observed with alcohol ethoxylates), improves wetting performance, and reduces foaming. The degree of peaking was also found to affect surface tension reduction. While a broad distribution yields low surface tension over a relatively broad range of ethylene oxide levels, a peaked methyl ester ethoxylate can provide even lower surface tension values, but only if the ethylene oxide content is optimized.

Another feature which differentiates methyl ester ethoxylates from alcohol ethoxylates is unsaturation. Methyl esters, particularly those in the tallow range, are relatively highly unsaturated, generally >50%. Alcohols, in contrast, are typically fully saturated. The impact of unsaturation on performance can be significant, and is addressed in the last section of this chapter.

As with all ethoxylates, the relationship between mols of and weight-percent of ethylene oxide is nonlinear for methyl ester ethoxylates. This relationship is also slightly different from that for the corresponding alcohol ethoxylate because of the difference in molecular weight between the feedstocks. The relationship between mols and weight-percent ethylene oxide for various methyl ester ethoxylates is shown in Figure 5.

IV. FORMULATING DETERGENTS WITH METHYL ESTER ETHOXYLATES

A. Water Solubility

The relationship between water solubility and the degree of ethoxylation (polyethylene oxide chain length) is well understood for alcohol ethoxylates. When the degree of ethoxylation is expressed as average weight-percent, an alcohol

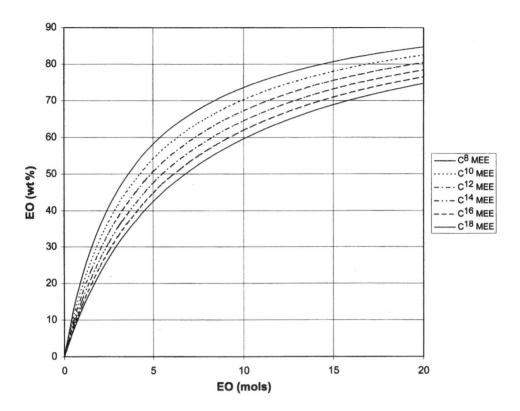

FIG. 5 Relationship between moles EO and wt% EO for C_8, C_{10}, C_{12}, C_{14}, C_{16}, and C_{18} methyl ester ethoxylate:

$$\text{Wt\%} = \frac{(\text{mols EO}) \, (44) \times 100}{[(\text{mols EO}) \, (44) + (\text{mol. wt. of ME})]}$$

ethoxylate is generally water soluble if the ethylene oxide content averages > 50%. At an average ethylene oxide content of 50%, alcohol ethoxylates are generally considered borderline water soluble.

Methyl ester ethoxylates, however, are inherently less water soluble than their alcohol ethoxylate counterparts because they contain a terminal methoxy group in place of the more hydrophilic hydroxyl group. Consequently, it takes a higher degree of ethoxylation for methyl ester ethoxylates to achieve water solubility. Although it is difficult to determine the precise ethoxylation level needed to achieve water solubility, studies suggest that methyl ester ethoxylates require at least 55 weight-percent of ethylene oxide to achieve borderline water solubility. Addition of more ethylene oxide to the polyethylene oxide chain logically in-

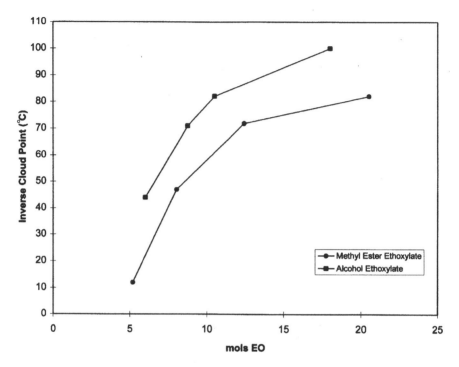

FIG. 6 Water solubility (inverse cloud point temperature) as a function of EO content (mols) for C_{12-16} alcohol and methyl ester ethoxylates made with Ca/Al-alkoxide Catalyst. (Cloud point temperature = temperature at which 1% aqueous solution turns cloudy upon slow heating; C_{12-16} alcohol ethoxylate is described in Table 1; distribution of methyl ester = 9% C_8, 8% C_{10}, 46% C_{12}, 18% C_{14}, 9% C_{16}, and 10% C_{18}.) (From Ref. 23.)

creases water solubility. Each mole of ethylene oxide has less of an impact on solubility than the alcohol ethoxylates, as illustrated in Figure 6.

Figure 6 compares the inverse cloud point temperatures (temperature at which 1% aqueous ethoxylate solutions cloud, as temperature is increased) as a function of average ethylene oxide content for C_{12-16} alcohol ethoxylates and C_{12-16} methyl ester ethoxylates. Methyl ester ethoxylates generally give inverse cloud point temperatures that are roughly 10°C lower than those their alcohol ethoxylate counterparts. Although the desired inverse cloud point temperature can be obtained by adding additional ethylene oxide to the methyl ester, the amount of ethylene oxide required depends on the target inverse cloud point. For example, to achieve a cloud point of about 45°C, a C_{12-16} methyl ester ethoxylate requires about 2 additional mols of ethylene oxide compared to a C_{12-16} linear alcohol ethoxylate. To obtain an inverse cloud point of 80°C, approximately 8 additional mols of ethylene oxide would need to be added.

In practice, inverse cloud point temperature is used more as a quality control measure rather than as a solubility requirement. Although methyl ester ethoxylates are less water soluble than their alcohol ethoxylate counterparts, they can achieve comparable formulatability characteristics at somewhat higher ethylene oxide levels.

Water solubility is affected by carbon chain length. A shorter carbon chain length results in greater water solubility (higher inverse cloud point temperatures) unless the level of ethylene oxide is unusually high [14,16]. At high ethylene oxide levels, the impact of carbon chain length is effectively nullified because solubility is almost entirely controlled by the long polyethylene oxide chain.

B. Viscosity/Gel Formation

Alcohol ethoxylates tend to form gels in the preparation of aqueous solutions. These gels readily form as ethoxylate concentration, average ethylene oxide content, and ionic strength are increased and as temperature is decreased. To destroy them can require substantial mechanical mixing and sometimes heat.

Methyl ester ethoxylates have been shown to have a significantly reduced tendency to form gels [9,16]. For example, the viscosity behavior of a commonly used 7-mole lauryl-range ethoxylate (based on a modified Oxo-type C_{12-15} alcohol) is illustrated in Figure 7. Gross viscosity was examined as a function of concentration and ionic strength (sodium chloride concentration) at 10, 25, and 40°C. As illustrated, gels (shown in black in Fig. 7) form readily, particularly at room temperature and below.

Figure 8 shows the gross viscosity behavior for a comparable methyl ester ethoxylate. The lesser tendency of methyl ester ethoxylate to form highly viscous solutions and gels is related to structural difference between the two ethoxylates. Methyl ester ethoxylates lack a terminal hydroxyl group which can participate in forming more structured liquids via hydrogen bonding. Methyl ester ethoxylates also have an ester linkage which increases steric constraint and reduces the ability of the surfactant to form complex structures in solution.

For formulating liquid products, absence of gel formation is an advantage. Since methyl ester ethoxylates do not normally require passing through a gel stage, they dissolve more rapidly. Methyl ester ethoxylates may also be useful in reducing the gelling of aqueous solution of other ethoxylates, such as alcohol ethoxylates and alcohol ether sulfates.

Even at a lessened tendency to gel formation, methyl ester ethoxylates are influenced by the same compositional factors as other ethoxylates. As shown in Figures 9 and 10, increasing carbon chain length and ethylene oxide content increase their tendency to form gels.

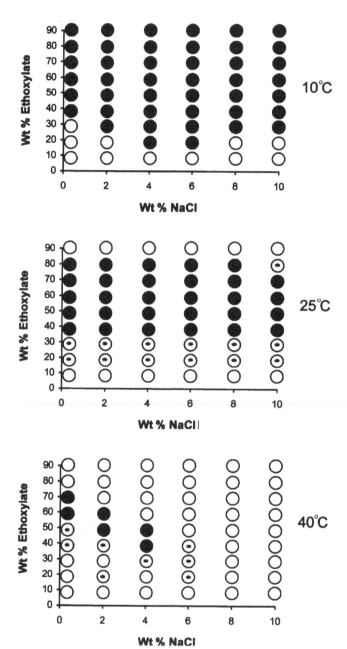

FIG. 7   Gel boundary diagram for C_{12-15} 7-mol modified-oxo-type alcohol ethoxylate at 10, 25, and 40°C. ($\circ$ = low viscosity liquid; $\odot$ = high viscosity liquid; $\bullet$ = gel.)

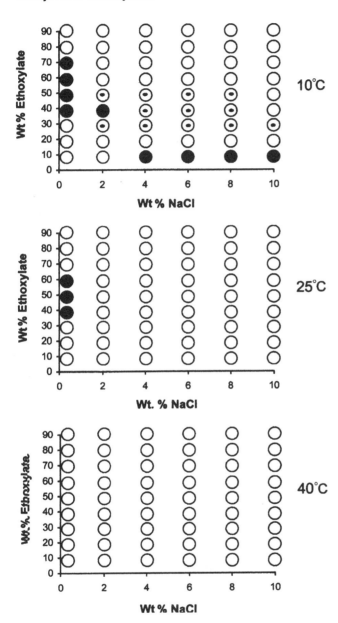

FIG. 8 Gel boundary diagrams for C_{12-16} 7.5-mol average methyl ester ethoxylate prepared using Ca/Al-alkoxide catalyst at 10, 25, and 40°C. (Distribution of methyl ester = 9% C_8, 8% C_{10}, 46% C_{12}, 18% C_{14}, 9% C_{16}, and 10% C_{18}. (○ = low viscosity liquid; ⊙ = high viscosity liquid; ● = gel.) (From Ref. 23.)

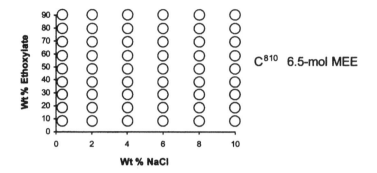

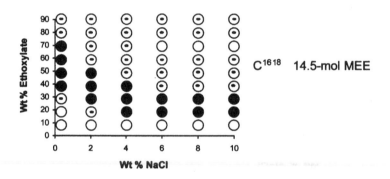

FIG. 9 Gel boundary diagrams for C_{8-10} and C_{12-16} methyl ester ethoxylates prepared using Ca/Al-alkoxide catalyst at 25°C. (Distribution of C_{8-10} methyl ester = 4% C_6, 53% C_8, and 43% C_{10}; distribution of C_{12-16} methyl ester = 57% C_{12}, 23% C_{14}, 12% C_{16}, and 8% C_{18}.) (○ = low viscosity liquid; ☉ = high viscosity liquid; ● = gel.) (From Ref. 23.)

C. Chemical Stability

The presence of an ester linkage makes methyl ester ethoxylates susceptible to base hydrolysis unlike their alcohol ethoxylate counterparts. Studies reported elsewhere concluded that aqueous solutions of methyl ester ethoxylates degraded with time and temperature at pH > 9 [9,15]. Studies examining the potential for hydrolysis occurring when methyl ester ethoxylates are included in powder formulations containing alkaline ingredients have yet to be reported.

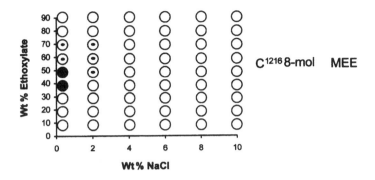

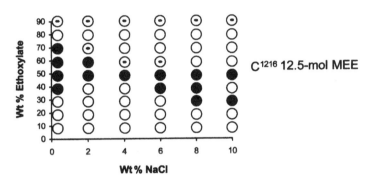

FIG. 10 Gel boundary diagrams for C_{12-16} 8-mol and 12.5-mol methyl ester ethoxylates prepared using Ca/Al-alkoxide catalyst at 25°C. (Distribution of C_{12-16} methyl ester = 57% C_{12}, 23% C_{14}, 12% C_{16}, and 8% C_{18}. (○ = low viscosity liquid; ⊙ = high viscosity liquid; ● = gel.) (From Ref. 23.)

D. Odor

As expected, methyl ester ethoxylates smell "different" from their conventional alcohol ethoxylate counterparts. Methyl ester ethoxylates made with short-chain (C_{8-10}) feedstocks have a clear advantage over conventional alcohol ethoxylates in both the level and the offensiveness of the odor. With short-chain ethoxylates, odor is generally determined by the level and odor of the unethoxylated feedstock. Short-chain (C_{8-10}) methyl ester is significantly less pungent than short-chain alcohol.

The opposite trend is true to a lesser degree with tallow-range ethoxylates. In general, ethoxylates of tallow-range alcohol have essentially no odor, while tallow-range methyl ester ethoxylates have a slight "methyl ester" odor.

The level of odor in the ethoxylates depends on carbon chain length and the degree of ethoxylation. As with alcohol ethoxylates, longer carbon chain lengths and higher ethylene oxide levels result in an overall reduction of odor level.

V. PERFORMANCE

A. Surface Properties

Methyl ester ethoxylates and their alcohol ethoxylate counterparts have similar surface properties. Gibbs' plot for pure C_{14} 7-mol (no other ethoxymers except the 7-mol homolog) methyl ester ethoxylate is compared to its pure C_{14} 7-mol alcohol ethoxylate counterpart in Figure 11. The methyl ester ethoxylate shows a higher critical micelle concentration (CMC) and a lower surface tension at the CMC than its alcohol ethoxylate equivalent. This increase in CMC is presum-

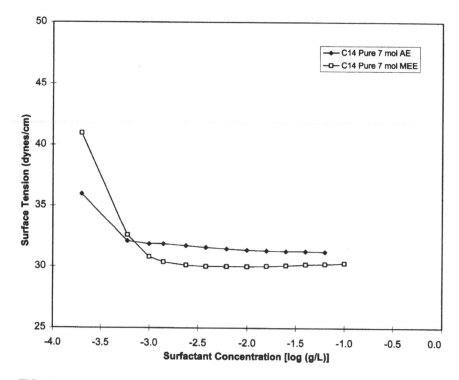

FIG. 11 Gibbs' plot for C_{14} pure (ethoxylates are pure 7-mol ethoxylates; there are no other homologs) 7-mol methyl ester and alcohol ethoxylates.

ably due to the presence of the ester moiety which adds rigidity and steric constraint to the methyl ester ethoxylate molecule. This would likely reduce the tendency of the molecule to micellize, leading to a slightly higher CMC.

The slightly lower surface tension of the methyl ester ethoxylate at and beyond the CMC is thought to be related to the presence of the terminal methoxy group on the ethylene oxide chain. In comparison to a terminal hydroxyl group, a terminal methoxy group is less hydrophilic. The resulting lower water solubility effectively drives more of the surfactant to the air/water interface, resulting in a lower surface tension.

At concentrations below the CMC, a longer carbon chain length yields a more efficient methyl ester ethoxylate surfactant [16]. At concentrations above the CMC, all carbon chain lengths are effective in lowering the surface tension once the ethylene oxide content is sufficiently optimized. Results also show that a methyl ester ethoxylate with a broader carbon chain length distribution is less surface active than one based on a more narrow-cut methyl ester.

The optimum level of ethylene oxide will vary depending on carbon chain length. In general, a lower degree of ethoxylation yields better surface properties, down to the limit where water solubility is no longer obtained.

There is also a difference in dynamic surface properties between methyl ester ethoxylates and alcohol ethoxylates. As shown in Figure 12 for pure 7-mole homologs, the methyl ester ethoxylate maintains a lower surface tension than its al-

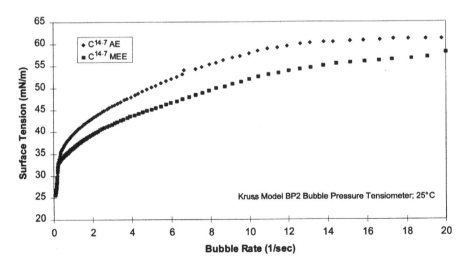

FIG. 12 Surface tension (nM/m) vs. bubble rate for C_{14} pure 7-mol methyl ester and alcohol ethoxylates. (Ethoxylates are pure 7-mol ethoxylates; there are no other homologs.)

cohol ethoxylate counterpart as measurements become more "dynamic" (bubble rate of bubble tensiometer is increased). This suggests that methyl ester ethoxylate is more effective in lowering surface tension (can achieve the same surface tension reduction with a lower surfactant concentration at the interface) and/or it diffuses through aqueous solution at a faster rate.

The impact of molecular structure on the ability of methyl ester ethoxylates to lower interfacial tension has been examined elsewhere [14]. These studies show that the relative performance of methyl ester and alcohol ethoxylates and the impact of ethylene oxide content and distribution depend strongly on the composition of the soil used to determine interfacial tension.

B. Soil Removal from Fabric

Based on the surface property data discussed earlier, methyl ester ethoxylates would be expected to perform similarly to alcohol ethoxylates, which is indeed the case. As shown in Figure 13, methyl ester ethoxylates are comparable to al-

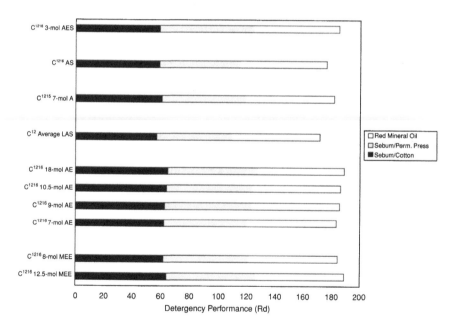

FIG. 13 Fabric detergency performance of methyl ester ethoxylate vs. conventional surfactants. (Conditions, test method, and protocol described elsewhere [15], descriptions of reference surfactants are given in Table 1; distribution for C_{12-16} methyl ester = 57% C_{12}, 23% C_{14}, 12% C_{16}, and 8% C_{18}; MEEs prepared using calcium/aluminum alkoxide catalyst. (From Ref. 23.)

TABLE 1 Reference Surfactants Used in MEE Assessment Studies

Abbreviation	Description
C_{12-16} 3-mol AES	C_{12-16}* 3-mol average sodium alcohol ether sulfate
C_{12-16} AS	C_{12-16}* sodium alcohol sulfate
C_{12-15} 7-mol AE	C_{12-15}† 7-mol average alcohol ethoxylate; alcohol "modified-oxo" type
C_{12} Average LAS	C_{12} average sodium linear alkylbenzene sulfonate (low 2-phenyl type)
C_{12-16} 7-, 9-, 10.5-, and 18-mol AE	C_{12-16}* alcohol ethoxylates

*Typical distribution of alcohol (linear, primary) = 68% C_{12}, 25% C_{14}, and 7% C_{16}.
†Typical distribution of alcohol = 20% C_{12}, 30% C_{13}, 30% C_{14}, and 20% C_{16}.

cohol ethoxylates, as well as other commonly used surfactants (see Table 1), in their ability to remove soil from fabric. Compositional variables affect the performance of methyl ester ethoxylates and alcohol ethoxylates. Lauryl-range and tallow-range methyl ester ethoxylates provide the best detergency, while optimum ethylene oxide content appears to depend on soil/cloth-type (see Fig. 14). Interaction with other surfactants also affects detergency performance. As shown in Figure 15, methyl ester ethoxylates act synergistically with alcohol ethoxylates, while the opposite can be observed with alcohol sulfate and alcohol ether sulfate.

C. Soil Removal from Hard Surfaces

Studies [24] have shown that linear alcohol ethoxylates made with short-chain alcohols are effective as hard-surface cleaners. A shorter-chain hydrophobe is thought to confer greater solvency on the surfactant, which assists in the removal of solid, greasy soils. An intermediate amount of ethylene oxide was also found to be generally best for hard-surface cleaning.

Previous studies [15,16] have shown that the same phenomenon is observed with methyl ester ethoxylates. Figure 16 indicates that, compared to conventional surfactants commonly used in hard-surface cleaners, methyl ester ethoxylates are highly effective. These results suggest that the increased solvency produced by reducing carbon chain length is magnified in some way with methyl ester ethoxylates. Whether or not the terminal hydroxyl group and/or the presence of the ester linkage produces this magnification is unknown.

Another way to demonstrate the impact of solvency on soil removal is to observe the change in soil weight upon immersion in aqueous surfactant solutions. Figure 17 shows how soil (lard) weight changes when immersed in 3% solutions

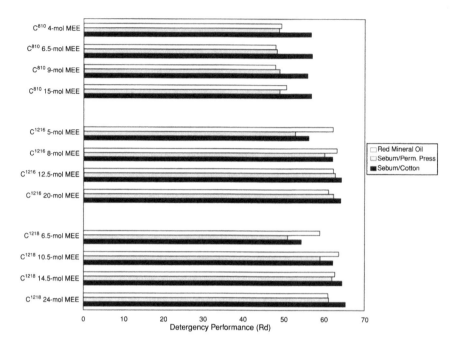

FIG. 14 Fabric detergency performance of methyl ester ethoxylates as a function of carbon chain length and EO content. (From Ref. 15.)

of various methyl ester ethoxylates. Short-chain (C_{8-10}) low-mole ethoxylates cause an increase in weight because they are effective in penetrating the soil during the immersion period. With higher carbon chain length and higher EO-containing methyl ester ethoxylates, soil weight is observed to decrease because the rate of soil removal (liquefaction, solubilization, emulsification) exceeds the rate of soil penetration.

Although short-chain ethoxylates are excellent surfactants for hard-surface cleaning, their performance is concentration dependent. With both alcohol ethoxylates and methyl ester ethoxylates, a shorter chain length improves hard-surface cleaning, particularly on greasy soils, but only at relatively high (> 1% surfactant) concentrations. At lower concentrations, performance is dictated less by solvency properties and more by surface properties. Consequently, a longer carbon chain length provides improved performance at low concentrations. The selection of the optimum performing ethoxylate therefore depends on use concentration as well as on soil and on substrate.

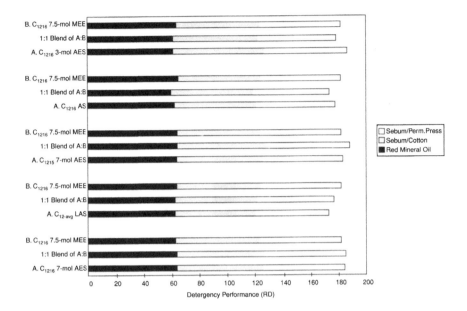

FIG. 15 Fabric detergency performance of C_{12-16} 7.5-mole methyl ester in 1:1 blends with other surfactants. (From Ref. 15.)

D. Foam Performance

Methyl ester ethoxylates are moderate foamers. They produce less foam than their alcohol ethoxylate counterparts (see Fig. 18) because they are sterically more constrained and because of the absence of a terminal hydroxyl group on the ethylene oxide chain.

Ethoxylate structure has a significant impact on foaming. As shown in Figure 19, increasing carbon chain length and ethylene oxide content results in higher foaming, which presumably correlates with the ability to lower surface tension. Hama has also shown that short-chain methyl ester ethoxylates, based on C_8 and C_{10} methyl esters, produce unstable foams [14].

The ability of methyl ester ethoxylates to produce foam appears to be sensitive to the presence of greasy soil. As shown in Figures 18 and 19, the addition of this soil significantly reduces foaming.

VI. APPLICATION OF METHYL ESTER ETHOXYLATES

Methyl ester ethoxylates are relatively new, and are just now being introduced into the market. In general, they can be considered to be interchangeable with al-

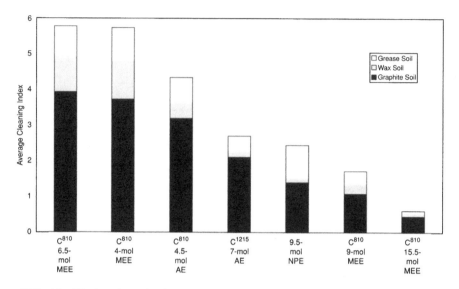

FIG. 16 Hard-surface cleaning performance of methyl ester ethoxylates vs. other surfactants. (From Ref. 15.)

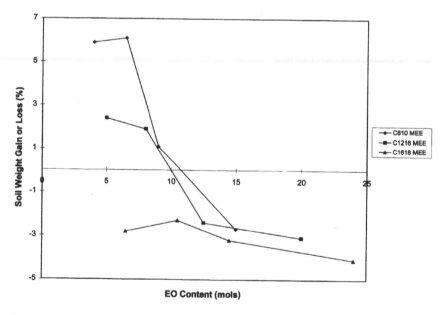

FIG. 17 Change in soil weight when soiled plates are dipped in aqueous solutions of methyl ester ethoxylates (1618 methyl ester = 31% C_{16} and 68% C_{18}). (From Ref. 15.)

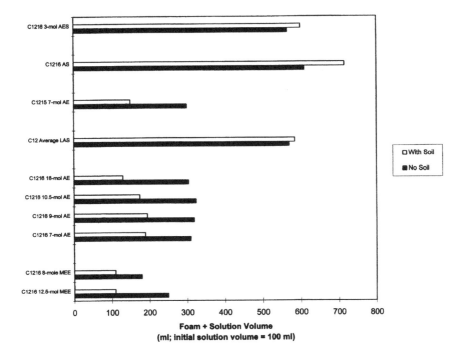

FIG. 18 Comparison of flash foaming of methyl ester ethoxylates vs. conventional sur-factants (100 ppm surfactant solutions; soil = 15% Crisco shortening, 15% olive oil, 15% instant mashed potato flakes, 30% milk, 25% water; test method (Schlag Foam Genera-tion) and protocol are described elsewhere [15]).

cohol ethoxylates except in terms of their pH stability, which limits their incorporation into aqueous formulations to a pH < 9. The following sections discuss the pros and cons for formulating methyl ester ethoxylates in detergents.

A. Laundry Powders

Methyl ester ethoxylates are good detergents, comparable with alcohol ethoxylates. They also appear to perform synergistically with other surfactants, which should be examined closely when developing finished formulations. They also foam less, making them a little more compatible for "controlled-foam" detergents. However, studies have yet to be published which discuss the ability of methyl ester ethoxylate to be processed into detergent powders. Methyl ester ethoxylates hydrolyze at high pH (> 9). Information regarding stability during

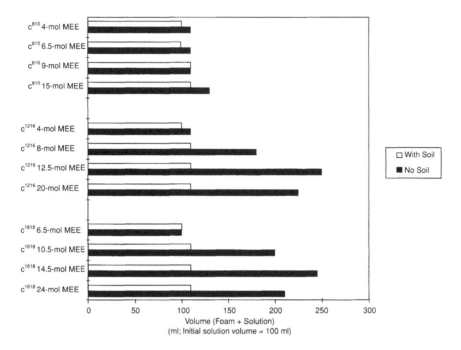

FIG. 19 Comparison of flash foaming of methyl ester ethoxylates as a function of carbon chain length and EO content (100 ppm surfactant solutions; soil = 15% Crisco shortening, 15% olive oil, 15% instant mashed potato flakes, 30% milk, 25% water; test method (Schlag Foam Generation) and protocol are described elsewhere [15]).

processing (spray-drying, agglomeration, etc.) and during storage is not yet available.

B. Laundry Liquids

Methyl ester ethoxylates should formulate well into liquids and provide acceptable stability provided pH remains < 9. Methyl ester ethoxylates should be easier to handle than their alcohol ethoxylate counterparts because of their reduced tendency to gel formation.

C. Hard-Surface Cleaners

Short-chain methyl ester ethoxylates appear to be outstanding detergents for removing solid soils from hard surfaces, but only when surfactant use concentration is significant (> 1%). At lower use concentrations, higher–carbon chain length methyl ester ethoxylates are more effective.

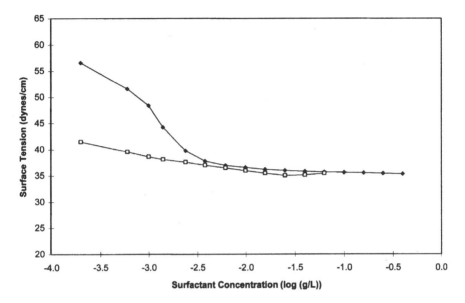

FIG. 20   Gibbs' plots of hydrogenated methyl tallowate ethoxylate (14.5 mols) vs. $C_{16/18}$ 14.5-mol MEE. [-◆- $C_{16/18}$ 14.5-mol MEE; -□- $C_{16/18}$ 14.5-mol MEE (hydrogenated methyl ester).] (From Ref. 15.)

Short-chain methyl ester ethoxylates also have an odor advantage over conventional alcohol ethoxylates, which can be important, particularly in household products. Formulation pH must be controlled to prevent hydrolysis of the methyl ester ethoxylates.

D.   Dishwashing Detergents

Since methyl ester ethoxylates are moderate foamers, and undergo significant hydrolysis > pH 9, they will not likely be used as the main surfactant in either hand or machine dish detergents. However, because of their ability to remove solid soil, methyl ester ethoxylates may find use not as foam stabilizers or foam-generating surfactants, but for enhancement of soil removal properties.

VII.   ETHOXYLATION OF OTHER ESTERS

Other esters (triglycerides, alkyl esters, fatty-fatty diesters, etc.) are also reasonable ethoxylation feedstocks [6], and are currently under study.

VIII. PROPOXYLATION OF METHYL ESTERS

Propoxylation, butoxylation, etc., of methyl ester ethoxylates is chemically straight-forward. Propoxylation of methyl esters is discussed in detail elsewhere [17].

IX. IMPACT OF UNSATURATION

The impact of unsaturation stems from the increase in rigidity caused by the presence of one or more double bonds to the alkyl chain. Although studies so far show that unsaturation has a relatively low impact on water solubility and viscosity, it has been reported to lower melting points by about 5 to 10°C [16]. Studies also show (Fig. 20) that unsaturation affects surface properties. Gibbs' plots of 14.5-mole ethoxylates produced from hydrogenated and nonhydrogenated C_{16-18} methyl ester show that unsaturation appears to increase surface tension below the critical micelle concentration. This suggests that unsaturation reduces the hydrophobic character of the methyl ester chain.

REFERENCES

1. MF Cox. In: A Cahn, ed. Proceedings of the 3rd World Conference on Detergents: Global Perspectives. Champaign, IL: AOCS Press, 1994:141.
2. HJ Scholz, H Suehler, JM Quack, W Schuler, M Trautmann. European patent application 89105357.1 to Hoechst AG (1989).
3. A Behler, HC Raths, K Friedrich, K Herrmann. German patent 39 14131 to Henkel KGaA (1990).
4. U Weerasooriya, CL Aeschbacher, BE Leach, J Lin, DT Robertson. U.S. patent 5,220,046 to Vista Chemical Company (1993).
5. F Yuji, H Itsuo, N Yuichi. U.S. patent 5,374,750 to Lion Corp. (1994).
6. U Weerasooriya, CL Aeschbacher, BE Leach, J Lin, DT Robertson. U.S. patent 5,386,045 to Vista Chemical Company (1995).
7. I Hama. INFORM 8(6):628–636, 1997.
8. T Tanaka, T Imanaka, T Kawaguchi, H Nagumo. European patent 0 783 012 A1 to Kao Corporation (1997).
9. I Hama, T Okamoto, H Nakamura. J Am Oil Chem Soc 72:781–784, 1995.
10. A Behler, H-C Raths, B Guckenbiehl. Tenside Surf Det 33:64–68, 1996.
11. T Imanaka, H Nagumo, T Tanaka, T Kono. Japanese patent JP 08323200 A2 to KAO Corp. (1997).
12. I Hama, H Sasamoto, T Okamoto. J Am Oil Chem Soc 74:817–822, 1997.
13. I Hama, M Sakaki, H Sasamoto. J Am Oil Chem Soc 74:829–835, 1997.
14. I Hama, M Sakaki, H Sasamoto. J Am Oil Chem Soc 74:823–827, 1997.
15. MF Cox, U Weerasooriya. J Am Oil Chem Soc 74:847–859, 1997.
16. MF Cox, U Weerasooriya. J Surfact Deterg 1:11–21, 1998.
17. MF Cox, U Weerasooriya, PA Filler, WH Mellors. J Surfact Deterg 1:167–175, 1998.
18. I Hama, T Okamoto, E Hidai, K Yamada. J Am Oil Chem Soc 74:19–24, 1997.

19. Y Sela, N Garti, S Magdassi. J Dispers Sci Technol 14(2):237–247, 1993.
20. R Subriana Pi, J Llosas Bigorra. German patent DE 196 11 508 C1 to Henkel KGaA (1997).
21. A Behler, A Folge. German patent DE 19 611 999 C1 to Henkel KGaA (1997).
22. K Kosswig. Tenside Surf Det 33(2):96–100, 1996.
23. BE Leach, ML Shannon, DL Wharry. U.S. patent 4,775,653 to Vista Chemical Company (1988).
24. MF Cox, TP Matson. J Am Oil Chem Soc 61:1273–1278, 1984.

6

N-Acyl ED3A Chelating Surfactants: Properties and Applications in Detergency

JOE CRUDDEN Hampshire Chemical Corporation, Nashua, New Hampshire

I. INTRODUCTION

The two fundamental functions of detergent systems are chelation of water hardness ions such as magnesium and calcium ions which, as elucidated by the DLVO theory [1], precipitate and coagulate colloidal dispersions, such as dispersed proteinaceous soils or surfactant systems, and surfactancy for wetting and removal of soil. Historically soap was used in sufficient excess to redissolve coagulated soil. Later in the development of detergents it was found that polyphosphates acted as very efficient complexing agents for water hardness ions and would strongly enhance the detergency of systems containing soaps or synthetic detergents. Unfortunately, in the 1960s the presence of phosphate builders in detergents was linked to eutrophication in inland waterways where the low availability of phosphate in the water had been acting as a growth-limiting factor for algae and other aquatic organisms. Ethylenediaminetetraacetate, EDTA, the very efficient chelate [2], was never considered a viable candidate because of very low rates of biodegradation. Nitrilotriacetic acid [2], NTA, was briefly hailed as the ideal builder since it acted as an extremely efficient complexing agent and biodegraded extremely rapidly. However, safety concerns, real or imagined, led to curtailment of its use in the United States. Subsequently zeolites became the builder of choice. Zeolites, unfortunately, are very insoluble and, as well as depositing gradually in fabrics, are causing increasing problems by building up in sewage treatment plants.

The quest continues unabated for the perfect builder and for more efficient, safer surfactant systems. Recent developments in the techniques of synthesis with hydrogen cyanide allow the production of multifunctional products which combine the properties of a strong and gentle surfactant with an efficient builder and complexing agent. These products should contribute significantly to the development of safer, more effective, and more versatile detergent systems.

II. STRUCTURE AND SYNTHESIS OF N-ACYL ED3A CHELATING SURFACTANTS

EDTA (Fig. 1a) is a well-known versatile chelating agent which forms soluble complexes with transition metal and water hardness ions. When it forms a chelate complex with most metal ions only three of the four carboxylate groups participate in the chelation process. The fourth acetate group remains pendent in solution and exerts little influence on the chelation process. The point of attachment of this fourth acetate group provides an ideal site of attachment for a functional group which could provide enhanced functionality to the molecule. However, since EDTA is symmetrical, selective replacement of just one acetate group in high yield is not possible directly, and replacement of more than one will severely compromise the power of the chelate.

Parker [3–9] developed an indirect strategy in which he first produced ethylenediamenetriacetate, ED3A (Fig. 1b), in high yield and subsequently attached an acyl group to the vacant fourth position under carefully controlled conditions.

The structure of Lauroyl ED3A is depicted in Figure 1c. Under acidic conditions, below pH 4, the product exists in triacid form as fine white crystals which are insoluble in water and organic solvents. As the product is neutralized with a base such as sodium hydroxide or triethanolamine (Fig. 2), the water-soluble acetate salts begin to predominate and crystal-clear aqueous solutions form above about pH 5.

A wide range of N-acyl ED3A chelating surfactant salts can be produced by varying the acyl group and the neutralizing base. Acyl groups of C8 to C18 are commonly used to produce surfactants, but acyl groups below C8 will produce chelates of unique character. The choice of neutralizing base will allow further flexibility in modifying the products characteristics. Possible neutralizing bases include sodium hydroxide, potassium hydroxide, ammonium hydroxide, and amino alcohols such as triethanolamine or monoethanolamine. The triethanolammonium salts produce products of greater oil solubility or lower HLB than the corresponding sodium salts. A range of salt forms which can be produced and should provide a broad spectrum of properties is presented in Table 1.

N-acyl ED3A chelating surfactants when evaluated for surfactant and chelating properties exhibited a wide range of unique and unexpected features. Many of these properties are extremely desirable in detergent systems:

The products exhibit aquatic toxicity orders of magnitude lower than conventional anionics.
They are nontoxic to mammals.
They are essentially a nonirritant in comparison to conventional surfactants.
They are inherently biodegradable.

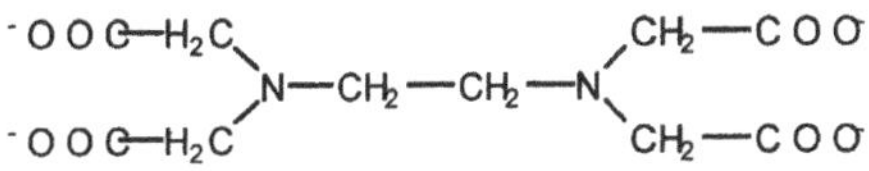

$$^-OOC-H_2C \diagdown \underset{^-OOC-H_2C} N-CH_2-CH_2-N \diagup CH_2-COO^- \underset{CH_2-COO^-} {}$$

(a)

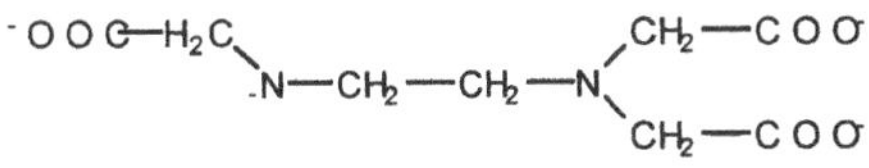

$$^-OOC-H_2C \diagdown \underset{^-OOC-H_2C} {:}N-CH_2-CH_2-N \diagup CH_2-COO^- \underset{CH_2-COO^-} {}$$

(b)

$$^-OOC-H_2C \diagdown \underset{O=C} N-CH_2-CH_2-N \diagup CH_2-COO^- \underset{CH_2-COO^-} {}$$

O=C
|
CH₂
|
CH₂
|
CH₂
|
CH₂
|
CH₂
|
CH₂
|
CH₂
|
CH₂
|
CH₂
|
CH₃

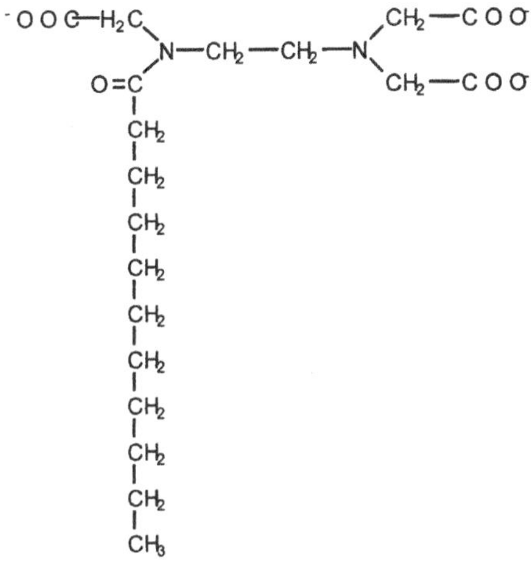

(c)

FIG. 1 (a) Ethylenediaminetetraacetate. (b) Ethylenediaminetriacetate. (c) Lauroyl ethylenediaminetriacetate.

They tolerate and are activated by water hardness ions.
They are soluble in high levels of caustic.
They act as extremely efficient hydrotropes.
They are compatible and synergistic with detergent enzyme systems.
They are compatible with cationic, amphoteric, anionic, and nonionic surfactants.
They passivate metals and act as corrosion inhibitors.

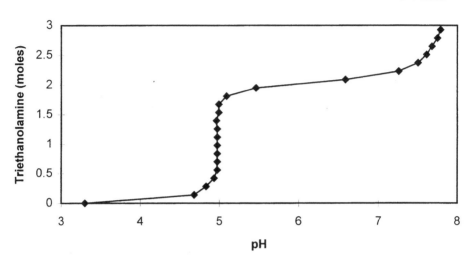

FIG. 2   Titration of LED3A with Triethanolamine.

TABLE 1   Known Salts of ED3A

Counterion	Acyl group							
	C_8	C_{10}	C_{12}	C_{14}	C_{16}	C_{18}	$C_{18=}$	Cocoyl
Sodium	*	*	*	*	*	*	*	*
Potassium	*	*	*	*	*	*	*	*
Ammonium	*	*	*	*	*	*	*	*
Monoethanolamine	*	*	*	*	*	*	*	*
Diethanolamine	*	*	*	*	*	*	*	*
Triethanolamine	*	*	*	*	*	*	*	*
N-Propylamine	*	*	*	*	*	*	*	*
Isopropylamine	*	*	*	*	*	*	*	*
2-Amino-1-butanol	*	*	*	*	*	*	*	*
2-Amino-2-methyl-1,3-propane diol	*	*	*	*	*	*	*	*
2-Amino-2-methyl-1-propanol	*	*	*	*	*	*	*	*
2-Amino-2-ethyl-1,3-propane diol	*	*	*	*	*	*	*	*
Tris(hydroxymethyl) aminomethane	*	*	*	*	*	*	*	*

The products should prove effective in detergent applications as diverse as laundry detergents, bottle washing, hard surface cleaning, ultramild baby shampoos, and possibly in oral and biomedical applications.

III. PHYSICAL PROPERTIES

A. Surfactant Properties

N-Lauroyl ED3A acid, below pH 4, is a white, insoluble crystalline powder. Above pH 5, whereupon two of the acetate groups become neutralized, crystal-clear aqueous solutions form. Since the molecule is amphiphilic in nature, it was expected that the product should exhibit surface activity. Two primary characteristics of a surfactant are depression of surface tension of aqueous solutions and the presence of a critical micelle concentration (CMC). Acetate salts of N-acyl ED3A were found to exhibit both of these properties.

The CMC and minimum surface tension for a range of these surfactants are presented in Table 2. Values for sodium lauryl sulfate are included for comparison. The CMC values are quite low, and minimum surface tension values, in the mid-20 dyne cm^{-1} range, qualify the products as quite effective surfactants. Like other aminocarboxylic surfactants, the surface tension of N-acyl ED3A is pH dependent even above the CMC.

The pH dependence of a range of salts of LED3A at a concentration of 1%, well above the CMC, is presented in Figure 3. Surface tension reduction is most highly developed between pH 5 and 7. Below pH 5 the acid form of the substance begins to separate from solution and the surface tension rises abruptly. Above pH 7 the triacetate salt is presumably the predominant form and the extremely high solubility of the headgroup in this form apparently reduces the surface activity.

B. Lather Stability

The potential of a surfactant to form stable, copious lather is important in many applications such as shampoos, shower gels, shaving creams, dishwashing liq-

TABLE 2 CMC and Minimum Surface Tension for Na N-Acyl ED3A at pH 7

Surfactant	CMC (w/w)% $\times 10^{-1}$	Minimum surface tension dynes cm^{-1}
Na Lauroyl ED3A	1.70	25.0
Na Myristoyl ED3A	0.27	21.5
Na Cocoyl ED3A	1.70	24.7
Na Oleoyl ED3A	0.99	28.0
Na Lauryl Sulfate	2.40	33.5

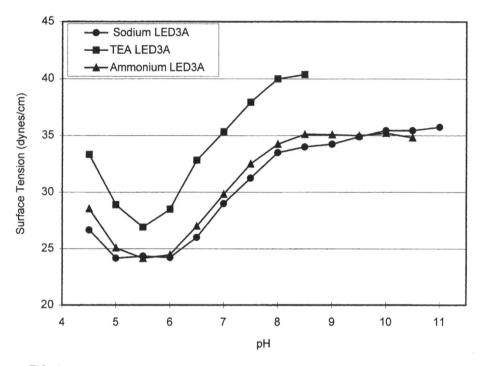

FIG. 3 Surface tension vs. pH (LED3A 1% solution).

uids, and foam cleansers for rugs and upholstery. In other areas, such as bottle washing, parts cleaning, spray adjuvants, and emulsion polymerization, the presence of copious stable foam can be detrimental. A useful method for the measurement of the propensity of surfactant systems to form stable lather under conditions of high shear was developed by Hart and DeGeorge [10]. This method was used to assess the stability of lather formed on solutions of LED3A and to assess the impact of added electrolytes on the stability. A product having a lather drainage time of < 20 sec at 2% activity would represent a fairly poor lathering agent and a product scoring > 60 sec would be rated as quite effective.

1. Lather Stability in the Presence of Sodium Chloride

The lather drainage time for a 1% solution of Na LED3A, in the presence of an increasing concentration of sodium chloride, was determinded by the method of Hart and DeGeorge [10]. The results are presented in Figure 4. Values for sodium lauryl sulfate are included for comparison. The lather stability on SLS is depressed by the addition of sodium chloride but, by contrast the lather on Na LED3A is strongly enhanced by the addition of sodium chloride reaching a lather drainage time at 4% sodium chloride which is twice that in the absence of added electrolyte.

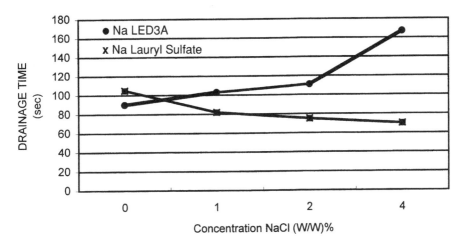

FIG. 4 Lather drainage time vs. salinity, for 1% surfactant solutions.

2. Lather Stability in the Presence of Water Hardness

Divalent water hardness ions such as magnesium and calcium rapidly delather most conventional anionic surfactants. This is a major reason for incorporating builders and complexing agents into detergent formulations. By contrast, the presence of water hardness significantly increases the lather drainage time of Na LED3A. The addition of 2000 ppm water hardness increases the lather drainage time by a factor of 3, from 100 to 300 sec. The same level of hardness depresses the lather on sodium lauryl sulfate by a factor of 3, from 100 to 30 sec. It is likely that the divalent counterions can form bridges between adjacent headgroups in the surface film which leads to the formation of a weblike stable structure at the interface.

3. Synergistic Lather Enhancement on Mixed Surfactant Systems

N-acyl ED3A chelating surfactants are capable of very dramatic, highly synergistic lather enhancement on mixed surfactant systems particularly in the presence of divalent counterions. The lather stability on mixtures of SLES (3 moles EO) and Na LED3A at a constant concentration of 1% in the presence and absence of water hardness is presented in Figure 5. Synergistic lather enhancement is evident when the system composition passes 30% LED3A.

When Na LED3A was added to a commercial baby shampoo (Fig. 6), the lather stability was enhanced nearly sevenfold in soft water and in water containing 200 ppm water hardness. The two factors which lead to the formation of stable lather are low surface tension and high surface viscosity. Since synergy in surface tension is not evident in the mixtures, it is likely that synergistic lather enhancement is due to increase of surface viscosity and formation of more stable surface films which are

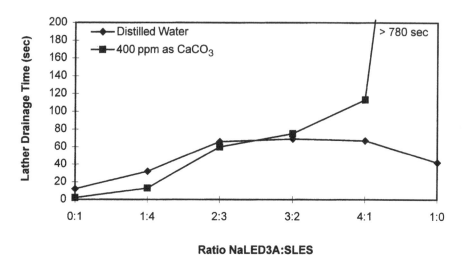

FIG. 5   Lather drainage time vs. composition, Na LED3A:SLES 1% surfactant.

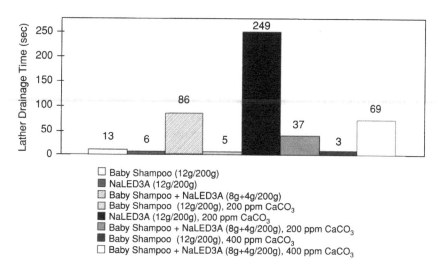

FIG. 6   Lather drainage time, Na LED3A and Baby Shampoo, 34°C, pH 6.48.

further enhanced by interlinking of the surfactant headgroups by the divalent counterions. Synergistic lather enhancement has also been noted on other systems [11].

4. Control of Foam on N-Acyl ED3A

In soft water, hydrocarbon oils and and silicone antifoams will defoam solutions of Na LED3A. If significant water hardness is present, control of foam becomes more difficult. Stearoyl ED3A produces stable copious lather in soft water, but is rapidly defoamed in the presence of water hardness. Sodium nonanoyl ED3A produces hardly any lather in the presence or absence of water hardness.

C. Chelating Properties

1. Calcium Chelation Value

Since N-acyl ED3A chelating surfactants retain the three acetate groups present on EDTA responsible for metal ion chelation, it was expected that the products would act as quite efficient chelates. Surprisingly, unlike conventional chelates, Na LED3A was found to exhibit chelating power which is concentration dependent. In other words, whereas EDTA will complex a fixed amount of calcium per gram of chelate no matter what the concentration, LED3A will complex more grams of calcium per gram of chelate, if the solution is more concentrated. Below its CMC, LED3A does not act as a very efficient chelate, but above the CMC chelation of calcium eventually approaches an efficiency of 1:1.

The calcium chelation value of a substance can be determined by an oxalate titration. In this method [12], a sample of the product to be tested is added to a sodium oxalate solution at pH 10 to 12. The mixture is then titrated with a solution containing calcium ions. When the chelate becomes depleted the endpoint can be noted by the precipitation of insoluble calcium oxalate. The calcium chelation value (CV) is reported as mg of calcium chelated per gram of chelate. The CV of Na LED3A was determined at a range of concentrations. The results are presented in Figure 7. EDTA chelates 260 mg $CaCO_3$/g irrespective of system concentration.

2. Order of Chelation of Metals

The order of chelation of metals was found from spectroscopic measurements to be:

$$Mg^{2+} < Cd^{2+} < Ni^{2+} \sim Cu^{2+} < Pb^{2+} \leq Fe^{3+}$$

IV. DETERGENT PROPERTIES

A. Influence of pH on the Detergency of LED3A

Since the surface tension of N-acyl ED3A solution is pH dependent and since detergency is often alkalinity dependent for other reasons, the dependence of the detergency of Na LED3A on solution pH was investigated by tergotometry (Terg-o-tometer model 7243S, United States Testing Company, Hoboken, NJ). The test fabric used was cotton uniformly stained with dust/sebum. The surfac-

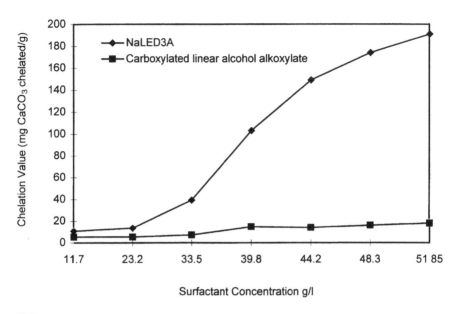

FIG. 7 Calcium chelation value.

tant concentration was 0.2 (w/w)%. The fabrics were washed in the Terg-o-tome-ter at 40°C for 15 min and rinsed for 5 min. Distilled deionized water was used and the pH was adjusted to a range of values between 5 and 12 using sodium hydroxide solution. Reflectance gain (Delta) was determined using a photovolt meter (model 577, Photovolt, Indianapolis) fitted with a detergent head and green filter. The results are presented in Figure 8. Under these conditions brightness is enhanced by increasing the pH of the system.

B. Detergency of Na LED3A:LAS Mixtures

Linear alkylbenzene sulfonate is one of the most commonly used surfactants in modern detergents. The surface tension of blends of varying proportions of Na LED3A and LAS, at a constant surfactant concentration of 1 (w/w)%, was determined using a Kruss K12 tensiometer fitted with a platinum Willhelmy plate. The results are presented in Figure 9. The mixture exhibits strong synergy in surface tension reduction which is most pronounced at a ratio of four parts LED3A to one part LAS.

The same systems were evaluated for detergency at a concentration of 0.2 (w/w)% in the Terg-o-tometer. The test fabric was cotton uniformly stained with dust/sebum. The test was carried at a temperature of 40°C at pH 10 in soft water. Wash time was 15 min and rinse time was 5 min. The results are presented in Figure 10. Na LED3A can be seen to be more effective than LAS under the con-

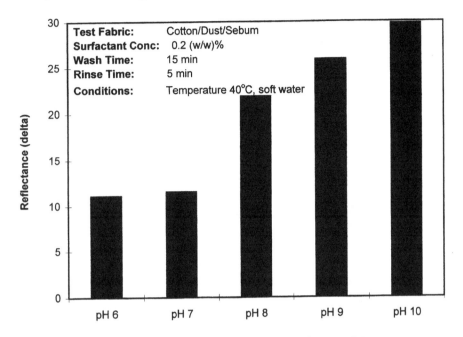

FIG. 8   Influence of pH on detergency of Na LED3A: (brightness gain).

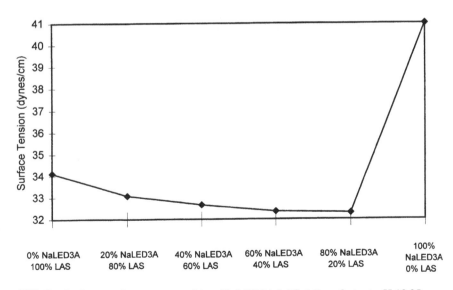

FIG. 9   Surface tension vs. composition, Na LED3A:LAS, 1% surfactant, pH 10.25.

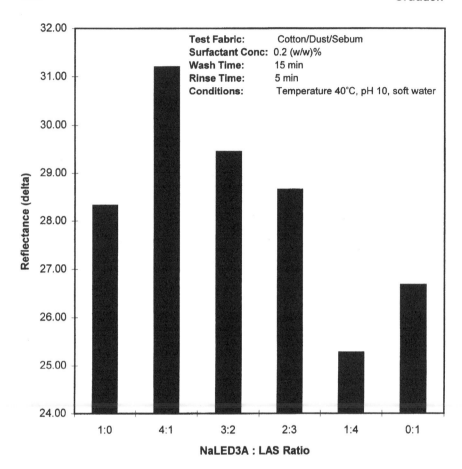

FIG. 10   Influence of composition on detergency.

ditions of the test, but a strong synergy is evident in the mixture containing four parts NaLED3A to one part LAS attaining a brightness gain more than 5 points higher than LAS alone at the same concentration.

C.   Detergency of APG600 and SLES with LED3A

Detergency of alkyl polyglycoside, Henkel APG600, and sodium lauryl ether sulfate, SLES, (3 moles EO) were evaluated in a similar manner (Fig. 11) alone and in combination with LED3A. LED3A significantly enhances the performance of both the SLES and the alkyl polyglycoside under the conditions of the test.

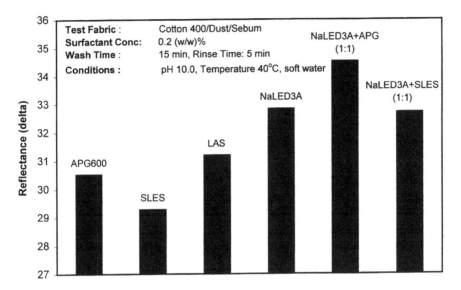

FIG. 11 Detergency of surfactant blends.

D. Draves Wetting Time

The Draves Wetting Time for a 0.1% solution of Na LED3A at pH 7 on cotton, nylon, and worsted wool is presented in Figure 12. The surfactant is quite efficient at wetting nylon and wool, but not cotton, under the conditions of the test. Mixtures of Na LED3A with lauryl dimethyl amine oxide and sodium lauroyl sarcosinate are more efficient at wetting cotton than the pure surfactant (Fig. 13).

E. Hydotrophic Activity of N-Acyl ED3A

A hydrotrope is an organic substance which increases the solubility in water of other organic substances. Most hydrotropes are not highly surface active and provide no performance benefit other than making a mixture homogenous. N-acyl ED3A was evaluated for its ability to clarify highly alkaline detergent systems which normally require the addition of a traditional hydrotrope like sodium xylene sulfonate.

Na LED3A was evaluated for its ability to clear three different surfactant systems—sodium dodecylbenzene sulfonate, C_{15} linear primary alcohol ethoxylate, and linear alkyl benzene sulfonate—in 5% sodium hydroxide. Two conventional hydrotropes, ammonium xylene sulfate and sodium alkanoate, were evaluated for comparison. The tests were carried out at 22°C and 70°C. The results are presented in Figures 14 and 15.

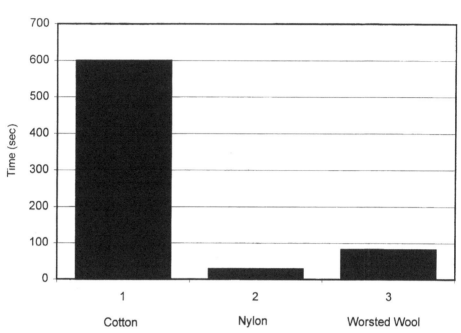

FIG. 12   Draves Wetting Times, 0.1% Na LED3A, pH 7, 23°C.

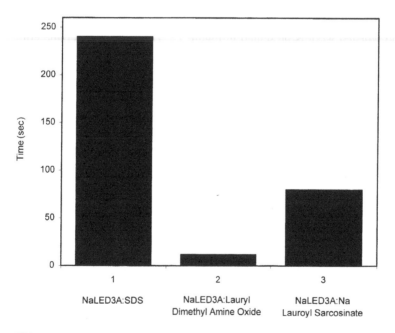

FIG. 13   Draves Wetting Times, cotton, pH 7, 23°C, 0.1% solution.

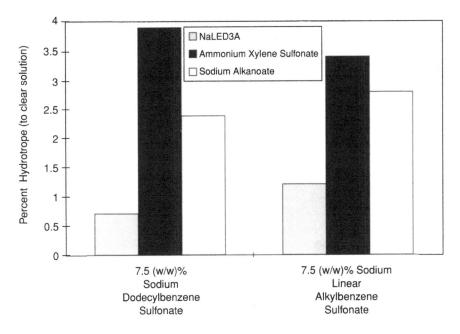

FIG. 14   Caustic system hydrotroping performance at 22°C.

It can be seen from Figure 14 that LED3A is more than twice as effective as sodium alkanoate and four times more effective than ammonium xylene sulfonate. LED3A was also evaluated as a substitute for conventional hydrotropes in the heavy-duty alkaline detergent formulation presented in Table 3.

The formulation in the absence of hydrotrope is a hazy two-phase system. A range of hydrotropes were evaluated for their ability to clear the system. The minimum quantity of hydrotrope required to produce a clear transparent single phase system at room temperature is presented in Figure 16. It is apparent from the data presented in Figure 16 that LED3A is a very effective hydrotrope in this system.

F.   Compatability of LED3A with Detergent Enzyme

In recent years detergent enzymes have become extremely important components in advanced detergent systems. The most common classes of enzyme are proteases, which help dissolve proteinaceous soils, and lipases, which help remove lipophilic soils. Enzymes, however, are delicate, complicated structures which are very susceptible to degradation and deactivation by aggressive proteolytic substances such as most anionic surfactants like sodium lauryl sulfate and linear alkyl benzene sulfate.

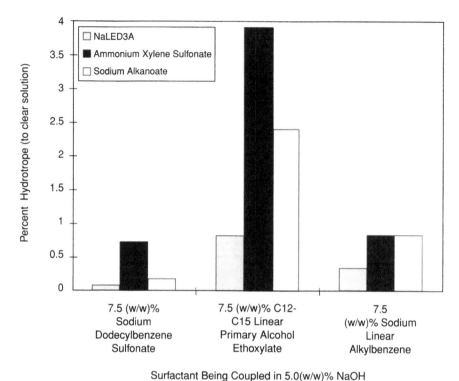

FIG. 15 Caustic system hydrotroping performance at 70°C.

TABLE 3 Heavy Duty-Alkaline Detergent

Sodium dodecylbenzene sulfonate (90%)	21.0
Triethanolamine	3.2
C_{12-15} Primary linear alcohol ethoxylate	5.2
NaOH (50%)	3.5
Water	67.1

A Terg-o-tometer study was carried out to determine the compatability of N-acyl ED3A chelating surfactants with Savinase, a protease enzyme [13]. Myristoyl and Oleoyl ED3A were neutralized with aqueous sodium hydroxide to produce 20 (w/w)% solutions and 10 g of the solutions were diluted to 1 L with distilled deionized water. The surfactant solution was then decanted into four of the Terg-o-tometer cells and allowed to equilibrate to 25°C. One milliliter of protease enzyme solution (Savinase 16.0L type EX) was diluted to 100 mL with distilled deionized

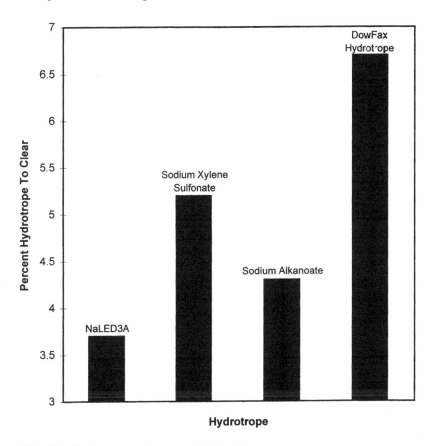

FIG. 16   Hydrotrope performance in HD alkaline detergent.

water and 1.43 mL of this enzyme solution was added to one of the cells containing each of the surfactants and allowed to acclimate for 20 min. Two swatches of cotton test fabric, uniformly soiled with blood/ink/milk (Scientific Services lot 5993 [14]), were placed in each cell and allowed to soak for 90 min. The Terg-o-tometer was activated and the swatches were washed for 30 min. The wash water was decanted and the swatches were rinsed for 10 min with clean water.

The swatches were removed from the cells and allowed to dry overnight. The brightness gain was determined using the Photovolt Meter, fitted with a detergent head and green filter. Four reflectance measurements, two for each side, were recorded for each swatch before and after washing. The brightness gain for each system is presented in Table 4.

It is evident from the data presented in Table 4 that these surfactants are compatible with the protease enzyme savinase. A similar study was carried out

TABLE 4 Brightness Gain for Surfactant: Enzyme
Combinations

System	Average brightness gain
Na MED3A	40.4
Na OED3A	43.4
Na MED3A + savinase	43.8
Na OED3A + savinase	45.3

TABLE 5 Brightness Gain for Surfactants and
Enzyme Alone and in Combination

System	Average brightness gain
Na LED3A	34.8
Na MED3A	33.7
Enzyme (savinase)	32.1
Na LED3A + enzyme	46.8
Na MED3A + enzyme	46.2

on Na Lauroyl ED3A and Na Myristoyl ED3A in which the cleaning power of
the enzyme was evaluated independently as well as in combination with the surfactants. The results are presented in Table 5.

It can be concluded that (1) this protease enzyme is compatible with N-acyl
ED3A; (2) the presence of N-acyl ED3A significantly enhances the cleaning
power of the enzyme solution; and (3) the presence of the protease significantly
enhances the cleaning power of the surfactant solution, contributing an extra 12
points of brightness.

G. Compatability with Activated Bleach Systems

TAED/perborate-activated bleach systems are used to provide low temperature
bleaching action in laundry detergents. The bleach system is particularly useful for
removing food dyes such as those present in tea or grape juice stains. Na LED3A
was evaluated for its ability to help remove grape juice stains from cotton/poly-
ester. The tests were carried out as before using the Terg-o-tometer at pH 7 and 9.
LAS and APG600 were also evaluated for comparison. The results are presented in
Figure 17. NaLED3A is the most efficient of the three surfactants at pH 9.

H. Corrosion Inhibition and Metal Passivation

N-acyl ED3A salts combine high surface activity with strong chelating power
and so are expected to adsorb tenaciously to the surface of metals. The ability of

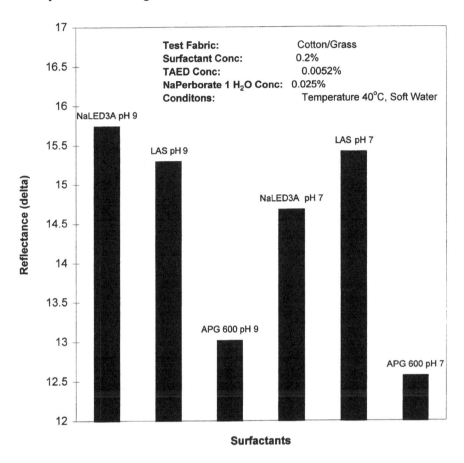

FIG. 17 Removal of grass stains with surfactant and activated bleach (brightness gain).

LED3A to inhibit the corrosion of bright mild steel was evaluated. Steel coupons were immersed in 15% HCl at 40°C for 96 hours. The progress of corrosion was followed gravimetrically. The data are presented in Figure 18. The control contained no corrosion inhibitor. The LED3A reduced the rate of corrosion by a factor of 10. Na LED3A has also been found to passivate stainless steel under highly alkaline conditions.

V. ENVIRONMENTAL PROPERTIES

A. Biodegradation

Before a new class of surfactant can be discharged to the environment it is essential to understand the potential environmental impact of the substance. Of primary

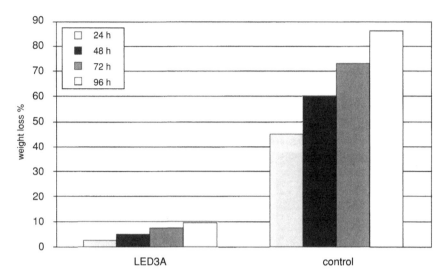

FIG. 18 Corrosion inhibition by sodium LED3A, dissolution of steel by 15% HCl at 40°C.

importance is the potential of microorganisms in the environment to break down the compound to simple harmless substances. In order to establish propensity of Na LED3A to biodegrade in an aerobic aqueous medium, an OECD Closed Bottle Test [15] was carried out. The test indicated that Lauroyl ED3A is "readily biodegradable" at a concentration of 2 mg/L, under the conditions of the test.

B. Microbial Growth Inhibition

The persistence of a substance in the environment depends on its degradation by the organisms present. If the substance reaches a concentration where it becomes toxic to these organisms, biodegradation will be impeded. An imbalance in the ecosystem could quickly result from rapidly increasing levels of the toxic substance as well as other substances which are no longer being biodegraded. The minimum inhibitory concentration (MIC) is the concentration of a test substance which just fails to inhibit microbial growth.

The Microbial Growth Inhibition Assay determines the toxicity of a substance to pure cultures of algae, bacteria, and fungi. Such a test carried out on LED3A [16] revealed that the substance failed to inhibit the growth of *Pseudomonas aeruginosa, Bacillus cereus, Glicocladium virens, Penicillium funiculosum, Aspergillus niger,* and *Oscillitoria prolifera* at concentrations up to 1000 ppm and only inhibited *Clostridium sporogenes* at concentrations above 500 ppm.

The results of the biodegradation study taken in combination with the microbial growth inhibition study indicate that it is extremely unlikely that LED 3A could reach levels in the environment where microbial ecosystems could become disrupted.

C. Acute Oral Toxicity

It is important to have a knowledge of the oral mammalian toxicity of a substance if widespread use is to be contemplated. An acute oral toxicity test was carried out on LED3A using Sprague-Dawley rats. The product was found to be nontoxic and to exert no toxic effects at 5 g/kg. Therefore an LD_{50} could not be established. The product can be defined as nontoxic under TSCA guidelines. By comparison, the LD_{50} of linear alkylbenzene sulfate has been reported [17] to be 900 mg/kg.

D. Aquatic Toxicity

The aquatic toxicity of Na Lauroyl ED3A to a range of aquatic organisms along the food chain was tested in compliance with TSCA guidelines. The organisms used were Rainbow Trout, Daphnia Magna, and the blue-green alga Selanastrum Capricornium Printz.

The substance was found to be nontoxic to Rainbow Trout at concentrations up to 320 mg/L during a 96-hour test. The acute toxicity LC_{50} (96 hour) is therefore > 360 mg/L. By comparison, the LC_{50} (96 hour) for LAS to Rainbow Trout [18] has been reported to be 0.36 mg/L

The aquatic toxicity of Na Lauroyl ED3A to Daphnia Magna was determined at concentrations up to 110 mg/L. No toxic or abnormal or sublethal effects were observed up to 110 mg/L. The acute toxicity, LC_{50}, is therefore > 110 mg/L. By comparison, the LC_{50} for LAS has been reported [19] to be 3.94 mg/L.

The aquatic toxicity, EC_{50}, of Na LED3A to Selenastrum Capricornium Printz was determined to be 1100 mg/L. The LC_{50} of LAS to this species has been reported [20] as being ~ 50 to 100 mg/L.

The aquatic toxicity studies carried out on Na LED3A have revealed that it exhibits aquatic toxicity which is orders of magnitude lower than conventional anionic surfactants.

E. Irritancy

Because it is envisaged that N-acyl ED3A chelating surfactants may be used in personal care products and since surfactants used in almost all applications come into close contact with humans and animals, extensive in vivo and in vitro irritancy studies were carried out [16].

1. Dermal Irritation

Na LED3A was found to cause no irritation to the skin of New Zealand white rabbits at a concentration of 1%.

2. Ocular Irritation

Na LED3A was determined to be a nonirritant to the eyes of albino rabbits at a concentration of 1%.

3. In Vitro Skin and Eye Irritation Assessment

Techniques developed by Advanced Tissue Sciences, La Jolla, CA [21], have allowed accurate assessment of skin and eye irritation potential of nominally irritating substances such as surfactants by studying their effects on the viability of live cultured human tissue. The methods are rapid and sensitive, and greatly reduce the need to use live test animals.

4. In Vitro Dermal Irritation

The Skin^2 MTT cytotoxicity assay [21] determines the degree to which the test substances reduce the viability of the test tissue over a 24-hour period. The results of the test correlate well with historical data from in vivo skin irritation studies. The in vitro scoring classification is presented in Table 6.

A range of surfactants including a series of N-acyl ED3A chelating surfactants were tested and classified according to this classification. The results are presented in Table 7.

The test classifies all the N-acyl ED3A salts tested as non-irritants with the next-mildest substance tested being Na Lauroyl glutamate a surfactant generally recognized as being very mild with a score of 4200 and a classification of mild. Not surprisingly the Baby Shampoo was also rated as mild with a score of > 2000.

5. In Vitro Eye Irritation

Advanced Tissue Sciences Skin^2 Cultures (model ZK 12,000) contain stomal and epithelial components which behave as in vitro counterparts of the cornea and conjunctivae, structures in the eye which are important targets in ocular irrita-

TABLE 6 In Vitro Scoring Classification

In-Vitro Score MTT-50 (µg/mL)	Classification
0–200	Severe
200–1,000	Moderate
1000–10,000	Mild
>10,000	Nonirritant

TABLE 7 In Vitro Irritancy Screen

Test compound	MTT-50 (µg/mL)	Classification
Na LED3A	>10,000	Nonirritant
Baby Shampoo	1,960	Mild
Baby Shampoo + Na LED3A 3:1	2,149	Mild
Na Lauroamphoacetate	628	Moderate
Na Laureth(3)sulfate	522	Moderate
Coco Betaine	560	Moderate
Sodium Lauryl Sulfate	280	Moderate
Na Lauroyl Glutamate	4,200	Mild
K LED3A	>10,000	Nonirritant
Na Nonanoyl ED3A	>10,000	Nonirritant

tion. Perkins et al. [22] have developed a technique, the Tissue Equivalent Assay, which uses this tissue to assess the potential of substances such as surfactants to act as eye irritants.

The concentrated test substance is applied topically to the tissue, and the time taken for cell viability to be reduced by 50%, the T_{50}, is determined by an MTT assay. The assay is capable of grouping test substances into the broad classifications—strong irritant, mild to moderate irritant, and slight irritant to innocuous. Test results have been found to correlate well with the results of historical in vivo studies. The scoring classification is presented in Table 8. Tissue Equivalent Assay test results for a range of test substances are presented in Table 9.

TABLE 8 Tissue Equivalent Assay
Scoring Classification

T_{50} (min)	Classification
<1	Strong irritant
>1–10	Mild to moderate irritant
>10	Slight irritant to innocuous

TABLE 9 Tissue Equivalent Assay: Eye Irritation Assessment

Test substance	T_{50} (min)	Classification
Na LED3A (10%)	>10	Slight irritant to innocuous
TEA LED3A (10%)	>10	Slight irritant to innocuous
Baby Shampoo (16% solids)	3.84	Mild to moderate irritant
Na LED3A (16% soln.)	5.65	Mild to moderate irritant

The results indicate that LED3A can be expected to cause very little eye irritation at a concentration of up to 10% and that at a concentration of 16% it is milder than baby shampoo at an equivalent concentration.

VI. CONCLUSION

N-acyl ED3A chelating surfactants are a new class of compound which act as strong and gentle surfactants and powerful complexing agents. They exhibit much lower aquatic toxicity than conventional anionic surfactants and are practically nonirritating and nontoxic to mammals. They are inherently biodegradable and as such are one of very few biodegradable chelates. Since chelation ability is concentration dependent, the products are very unlikely to mobilize heavy metal ions at discharge concentrations. The products are compatible with detergent enzymes and activated bleach systems and show synergistic lather enhancement with anionic surfactants, particularly in the presence of water hardness ions.

Since they are surface active chelates which adsorb strongly onto metals, they act as corrosion inhibitors for bright mild steel and passivate stainless steel under alkaline conditions. Given the broad spectrum of unique properties exhibited by these products, it is likely that they will find widespread use in a wide range of detergent and personal care applications.

REFERENCES

1. MJ Rosen. In: Surfactants and Interfacial Phenomena, 2nd ed. New York: John Wiley, 1989, pp 338–339.
2. Bersworth. U.S. patent 2,407,645.
3. U.S. patents 2,855,428 and 3,061,628, assigned to Hampshire Chemical Corp.
4. BA Parker, JJ Crudden. Proceedings, 4th World Surfactant Congress, Barcelona, Spain, 1996, Vol. 1, pp 446–460.
5. BA Parker. U.S. patent 5,177,243, assigned to Hampshire Chemical Corp.
6. BA Parker. U.S. patent 5,191,081, assigned to Hampshire Chemical Corp.
7. BA Parker. U.S. patent 5,191,106, assigned to Hampshire Chemical Corp.
8. BA Parker, BA Cullen, RR Gaudette. U.S. patent 5,250,728, assigned to Hampshire Chemical Corp.
9. BA Parker, BA Cullen. U.S. patent 5,284,972, assigned to Hampshire Chemical Corp.
10. JR Hart, MT DeGeorge. J Soc Cosmet Chem 31:223–236, 1980.
11. JJ Crudden, BA Parker, JV Lazzaro. Proceedings, IN-COSMETICS, Milan, Italy, February 1996.
12. Hampshire Chemical Corp., Data Sheet, Analytical Procedure for Calcium Chelation Value, HCT 16.
13. Savinase is Novo's trade name for protease enzyme. Novo Norkisk Biochem North America, Inc., Franklin, NC 27525.

14. Scientific Services S/D Inc., Sparrow Bush, NY 12780

15. Organization for Economic Cooperation and Development. OECD Method 301 D, "Closed Bottle Test."

16. JJ Crudden, BA Parker. Proceedings, 4th World Surfactant Congress, Barcelona, Spain, 1996, Vol. 3, pp 52–66.

17. EV Buchler, EA Neumann, WR King. Toxicol Appl Pharmacol 18:83–91, 1971

18. VM Brown, FSH Abram, LJ Collins. Tenside Detergents 15:57–59, 1978.

19. AW Maki, J Fish Res Board Can 36:411–421, 1979.

20. AN Yamane, M Okada, R Sudo. Water Res 18(9):1101–1105.

21. Advanced Tissue Sciences, The Lab Partner, Vol II, No. 7, July 1992.

22. MA Perkins, DA Roberts, R Osborne. Toxicologist 12:296, 1995.

7
Surfactants for the Prewash Market

MICHELLE M. WATTS Goldschmidt Chemical Corporation, Dublin, Ohio

I. INTRODUCTION

Cleaning of the body and clothes has been around for centuries. However, prewash soil and stain removers made their appearance in the 1960s. With the ban of phosphates in laundry detergents, clothes were not becoming as clean as they once were. The use of synthetic fibers also increased the need for a product to treat and remove stains.

The original non- or low-phosphate detergents were not very good at removing soils. Many soiled areas were difficult to remove with regular washing. Even with a second and a third washing, these stains were still present. Sometimes they could be removed if the garment was soaked in a solution of detergent. Hence, a need existed for a concentrated product to be placed on the piece of clothing before washing.

"Prespotters" are applied to stubborn stains, soils, and spots just prior to cleaning. With a short contact time, prespotters need to be highly effective. In order to be efficient, many of these products contain harsh surfactants and strong solvents.

A typical prespotter formulation would contain petroleum distillates, linear alcohol ethoxylates, and/or nonyl phenol ethoxylates. Potential problems with using a harsh product are flammability, low volatility, clothing damage by way of color removal, and skin irritation.

Numerous changes and improvements have been made in the detergent industry. Many of them have found their way into the prespotter formulations. This chapter will describe prewash technology and what commodity and specialty surfactants are used in this type of detergent.

II. BACKGROUND

In removing oil, grease, and stains from clothes, the use of a detergent aid is sometimes essential. What can be used to enhance the cleaning can fall into three categories: (1) enzymes; (2) bleach; and (3) prespotters, degreasers, etc.

Enzymes, used on protein-based stains, are good in hot water (45–55°C). However, they lose their efficiency with decrease in water temperature. Enzymes can be used in cold water if there is a lengthy presoak time before washing. However, they do not contribute significantly to washing efficacy in a 10-min cold (20°C) wash cycle [1,2].

In the prespot products that are aqueous based, enzymes can be found. The water-based prespotters are designed for protein- and vegetable-based pigment stains. The solvent-based products, in which enzymes are not soluble, are intended for removal of oil- and grease-based soil.

Bleaches can be used to pretreat clothes or added directly to the wash. Bleaches, both chlorine and nonchlorine, are effective for treating stains such as wine, fruit juices, etc., in hot water. Perborate bleaches are less effective in low-temperature wash. To improve the nonchlorine bleaches, a bleach activator is used.

Another problem with bleaches, especially chlorine bleach, is the damage to the fabric. If a piece of clothing has been washed constantly in a system with bleach, one may notice pinholing in the material.

Prespotter formulations may contain enzymes, surfactants, petroleum distillates, and other chemicals. With flammability and low volatility as an environmental issue, these products needed to be reformulated.

In this chapter, the words prewash, prespot, and pretreat will be used interchangeably.

A. Patent Literature

One of the earliest patents on prewash products is by Purex [3] in 1973. This patent describes a prewash composition that contains no water and does not gel. At that time, gelling of a prewash formulation was considered a negative attribute. The formulation contains a nonionic surfactant, an organic solvent, and an enzyme.

Over the last few years, different types of surfactants have found their way into the prewash formulations. Examples that have been mentioned in various patents [4–10] are listed below:

Ethoxylated secondary alcohols
Amine oxides
Quaternary ammonium compounds having one and/or two hydroxyethyl groups
Sodium dialkyl sulfosuccinate
Dicoco dimethyl ammonium chloride
Soaps
Monoalkyl trimethyl ammonium salt
Alkanolamides
Alkoxylated derivatives of ethylenediamine

Ethoxylated diamines
Ethoxylated amine oxides
Alkyl naphthalene sulfonates

B. Prewash Market

This market is very fragmented. There are major companies who manufacture pretreatment products along with many smaller companies. According to Chemical Marketing Reporter [11], a large section of the population (75%) pretreats their laundry. However, only one-third (33%) has purchased a prewash product, with the remainder using home solutions. Those who do not buy pretreatment products generally use their laundry detergent to spot treat stains.

Cleaning products, such as those used for hard-surface cleaning, have also been used for pretreating laundry. These products contain anionic and nonionic surfactants along with various cleaning solvents that are good for removing grease and grime.

The U.S. sales in the prewash category were ~$205 million for the year 1996. This represents ~5% of the detergent sales for the same time period [11]. The breakdown of types of products used can be found in Figure 1 [12].

III. PREWASH PRODUCTS

Products that can be used as prewash can vary. At times when a conventional pretreatment product is not very helpful, consumers have switched to products

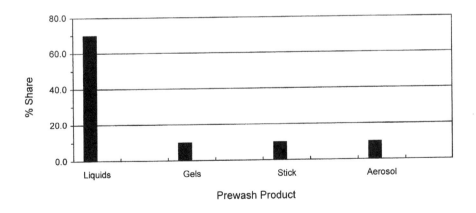

Compliments of SC Johnsom & Sons

FIG. 1 Prewash market segmentation. (From Ref. 12.)

that are normally used outside the laundry area. This market can then be divided into two categories:

1. Traditional—products that are sold as prespotters and detergent products
2. Nontraditional—products that are marketed for a nonlaundry application but have found use as a prespotter

A. Traditional Prespotters

Within the traditional area, the category can then be further separated into two subcategories: detergent products and pretreatment products. Since it has been established that less than half of consumers actually buy a pretreatment product, many consumers use their laundry detergent to treat their clothes. Instructions on the detergent packages inform the consumer on how to use the detergent as a prewash product.

Prespotters that are on the store shelves can be found in the following containers:

Squeeze tube
Spray or trigger bottle
Individualized prespotter packet
Stick
Tube with a plastic brush top
Aerosol

This represents a growth in product development over the years.

B. Nontraditional Prespotters

The nontraditional area includes products that are hard-surface cleaners, which could be rather mild, like window cleaners, or stronger formulations used on kitchen grease. These products contain the same types of surfactants, but also contain a solvent system. These products are normally used when there is an excessive amount of grease and grime.

Similar to what is found on the laundry package, many of the nonlaundry products suggest these would be suitable for removing stains on laundry. However, few consumers use these products as pretreatments.

IV. TESTING

Many ways exist to test the efficacy of prespotter products. In the laboratory, one can choose to use a Terg-o-tometer [13], Laundrometer [14], or even a washing machine. All three are valid. The common machine found in most laboratories is the Terg-o-tometer. With this instrument simulating a miniwashing machine, multiple testing can be done at one time.

The Laundrometer can also do multiple testing. However, this machine would simulate the horizontal axis washing machine which is popular in Europe. It will also be the standard washing machine in the future in the United States.

To properly evaluate a prespotter product or even a detergent product, the following items are important to consider:

Prespotter dosage
Detergent dosage
Water temperature
Water hardness
Types of soil/stains
Washing time
Fabric type

Two ASTM (American Society for Testing and Materials) procedures exist that discuss guidelines for cleaning soiled fabric [15,16]. The first one is for evaluating stain removal performance emphasizes that no single type of stain can predict overall performance of a product or treatment type. It is applicable to all types of laundry products, such as prespotters, detergents, bleaches, and detergent boosters. The suggestion is made that a variety of fabric types be examined. The removal of stains will be influenced by the fabric type. A minimum of six stains should be evaluated in order to properly characterize a stain removal product.

The second procedure is used for assessing the ability of detergent products to remove soils from fabric. The difference between this procedure and the preceding one is the type of soil used for testing. The only soil used in this test is standard soil.

If it is not feasible to test under more than one set of conditions, the following conditions which are found in Table 1 are recommended for use with a Terg-o-tometer:

An oily soil can be any of the following: butter, margarine, salad dressing,

TABLE 1 Testing Conditions

Water temperature	25–30°C (room temperature)
Water hardness	150 ppm
Washing time	10 min
Rinsing time	3 min
Liquid detergent	1.5 g/Liter (as-is basis)
Prespotter dosage	2 mL for liquids; 1 tsp for pastes
Conditioning time	5 min
Fabric types	Cotton, cotton-polyester
Soil types	Dust-sebum, standard soil, oily soils, various stains

% SR = [(L_w - L_u)/(L_n - L_u)] * 100

$$\% \ SR = [(L_w - L_u)/(L_n - L_u)] * 100$$

where:

L = reflectance

n = unstained unwashed fabric

w = stained washed fabric

u = stained unwashed fabric

FIG. 2 Percent soil removal (% SR).

gravy, etc. Standard soil can be obtained from TestFabrics Inc. (West Pittson, PA). The stains chosen, for example, coffee, tea, grass, mustard, etc., can be based on the experimenter's cleaning objectives.

Once washed, the swatches are read on a Hunter Lab Colorimeter for reflectance (L), redness/greenness (a), and yellowness/blueness (b) values. The reflectance is used to calculate percent soil removed (% SR) which ranges from 0 to 100. Whiteness index (W1) [17] and yellowness index (Y2) can also be determined. The whiteness scale ranges from 0 (black) to 100 (white). The yellowness index ranges from positive (yellow) to negative (blue). To determine how much soil is removed from the fabric, see Figure 2.

Whiteness and yellowness as indices will help in determining how well the prewash product or the individual surfactants are performing. A product or surfactant can be considered successful when these two criteria have been met: stain removal and whiteness. Many surfactants would remove the stain, but the tendency to yellow the fabric was great. The greater the tendency to yellow, the less white the fabric will be.

Testing of prewash products is time-consuming. Multiple testing should be done to obtain statistical data. With these data, the formulator can make claims as to the efficacy of the products.

V. CHEMISTRY

Surfactants can be classified in four ways: anionics, cationics, nonionics, and amphoterics.

Anionic surfactants are the primary ingredient in many types of cleaning products, including heavy-duty laundry detergents. Examples of these surfactants are sodium lauryl ether sulfate (SLES), sodium dodecyl benzene sulfonate

(Na DBS or LAS), and sodium alpha olefin sulfonate (AOS-Na). These products have high foaming capacity, good wetting agents, and good detergency. They are essential for removal of particulate soil.

Nonionic surfactants, which are also found in cleaning products, have been increasing their share of the detergent formulation. The most common nonionics used are linear alcohol ethoxylates (AE) and nonyl phenol ethoxylates (NPE). Some of the properties of nonionic surfactants are good detergency for oily soil removal, antiredeposition of soil for synthetic fabrics, low foaming, and good compatibility with anionic and cationic surfactants. Since many stains or spots on clothes are oily in type, nonionics are the largest ingredient in most prespotters. Food stains contain natural fats and oils while petroleum based stains are common on industrial clothing. NPE is generally somewhat more effective than AE.

Cationics generally do not show good detergency. However, from the patent literature [18,19], they are found in small quantities in detergents. Cationics are normally used as a fabric softeners and as a static-reducing agents.

Amphoterics are not normally found in detergents or prewash products because of their high cost. The benefits of using them include mildness, cleaning, wetting, enzyme compatibility, dye transfer inhibition, and foaming. Examples of amphoterics are cocamidopropyl betaine, sodium cocamphoproprionate, and cocamidopropyl hydroxysultaine.

A. Past Technology

The earlier commercial products that were sold in the supermarkets, drug stores, and mass merchandisers have many of the same surfactants. As shown in Table 2, the common surfactant found in the prewash products was ethoxylated alcohols, linear and nonylphenol.

Anionic and nonionic surfactants were used because they were, and still are, effective and economical. For specialty surfactants to have a role in prewash or detergent products, they need to show a cost benefit.

TABLE 2 Typical Composition of Commercial Prewash Products

Product	Anionic	Nonionic	Misc.
Commercial 1	Na DBS, SLS	—	—
Commercial 2	—	NPE, AE	Fatty acid Propylene glycol
Commercial 3	—	NPE, AE	—
Commercial 4	—	NPE, AE	Na citrate
Commercial 5	Oleyl methyl tauride	—	Na hydrosulfite
Commercial 6	—	NPE	—
Commercial 7	SXS	AE	—

B. Current Technology

As technology in the cleaning industry advances, so does the technology for prewash products. Examples of the current technology are products that can now be found as sticks and gels. New technology, as far as surfactants are concerned, is not much different from what was found in the literature 5 to 10 years ago. Nonionic surfactants are still being used. Enzymes are much more prevalent in formulations. Specialty surfactants are becoming part of the formulations.

Formulations appear to be moving from a product that can clean both water- and oil-based stains to cleaning very specific stains. These new pretreatments are becoming much more stain specific, such as:

Oily soils
Baby stains
Blood
Ink
Motor oil
Grass

C. Enzymes

Enzymes have been in detergent formulations since the 1960s [20]. A few years later, enymes made their appearance in the prewash patent literature [21,22]. At first, enzymes were only found in powdered or solid products. These products worked very well against protein-based stains. At that time, the enzymes could not be placed in aqueous formulations without decomposing. With advances made in enzyme synthesis technology, enzymes can now be placed in liquid detergent formulations and prespotters. Today, enzymes are widely found in prespot formulations.

D. Performance of Individual Surfactants

To develop new products for the prewash market, understanding how surfactants that are not normally found in detergent formulations perform as a prewash product is essential. In the past, the majority of the surfactants in these products were ethoxylated alcohols and anionic surfactants. These products are effective and economical.

Nontraditional detergent surfactants have been examined to enhance the performance of the prewash products. Surfactants, such as ethoxylated amines and amphoterics, have been examined. To understand the functions of these surfactants with respect to the conventional surfactants, individual testing should be done.

1. Testing of Surfactants

Screening of specialty surfactants as detergents is an ideal way to learn the strengths and weaknesses of these products. Testing individual surfactants will allow the formulator to relate structure versus cleaning performance. These specialty surfactants can give heightened performance in the proper formulation. Some products may not have great cleaning ability of dust-sebum and standard soil but may have exceptional capability to clean specific soils, such as grass, coffee, etc.

Surfactant testing was conducted under the following conditions:

150 ppm 40°C wash water
Wash time of 10 min; rinse time of 3 min
Cotton and cotton-polyester fabric
Dosage: 1.5 g/L–equivalent-actives basis
Stains—various

The following performance areas were considered:

Water hardness
Dye transfer
Synergistic cleaning performance
Cost performance benefit

2. Water Hardness

For a specialty surfactant to be fully beneficial, it needs to be able to clean in any water hardness. It is widely known that LAS loses its detergency when the water hardness increases. Various surfactants were examined under three water hardness conditions: 50 ppm (Ca/Mg ions), 150 ppm, and 300 ppm.

As was to be expected, decreasing the water hardness to 50 ppm aided the surfactants with improving their cleaning ability. Improvement was seen not only with soil removal but also with whiteness. Examples of this can be seen in Table 3. Only the cotton-polyester results are shown, but the same conclusions can be drawn with cotton fabric. Amphoterics showed good results on coffee and tea.

As the water hardness increases to 300 ppm, a decrease in cleaning was expected. A few specialty surfactants were oblivious to the hardness ions. The data in Table 4 are for cotton fabric. Using cotton-polyester will exhibit the same type of trends.

It is important for the specialty surfactant to be able to clean in any type of water hardness. Even though builders and sequestering agents are found in detergents to help the cleaning process by binding water hardness ions, these surfactants must add to and not detract from the cleaning process.

TABLE 3 Water Hardness Variable (cotton-polyester fabric)

Stain	Surfactant	Hardness (ppm)	% Soil removed	Whiteness index
Dust-sebum	PEG-30 glyceryl	50	49.3	52.0
	cocoate	150	45.5	46.8
Coffee	NaDBS	50	92.9	74.1
		150	86.5	64.7
Iced tea conc.	Disodium	50	72.8	72.7
	cocoampho diacetate	150	61.3	62.0
Coffee	Sodium	50	95.2	80.4
	cocoampho propionate	150	92.0	70.5
Grass	PEG-10	50	88.6	52.0
	coco amine	150	80.9	36.1

TABLE 4 Water Hardness Variable (cotton fabric)

Stain	Surfactant	Hardness (ppm)	% Soil removed	Whiteness index
Dust-sebum	PEG-30 glyceryl	150	56.7	42.7
	cocoate	300	62.5	43.8
Iced tea conc.	C12-15 alcohol ethoxylate (9)	150	53.9	42.1
		300	54.8	42.4
Grape juice	Disodium ricinoleamido MEA sulfosuccinate	150	68.0	79.8
		300	73.9	78.6
Coffee	Carboxylated NPE (10)	150	71.8	28.8
		300	75.2	32.6
Iced tea conc.	Disodium cocoampho diacetate	150	50.3	38.1
		300	50.9	40.0

3.   Dye Transfer Properties

With the arrival of the horizontal axis machine, dye transfer will become a more important issue to the consumer. With using less water for washing and rinsing, the detergent surfactants used will be in higher concentration. This is particularly true for prespotters where the typical consumer puts a large dose of surfactant on the spot or stain to be removed. Therefore, formulations need to clean without removing the dye from the fabric.

The most common product used to prevent dye transfer is PVP, polyvinyl pyrrolidone. This product is used for dye transfer inhibition and antiredeposition. PVP is not a surfactant, has no cleaning ability itself. The testing for dye transfer can be found in another ASTM method [23], which is summarized in Table 5. Delta E (total color difference) is presented as an average of three different color values. The lower the number, the more desirable.

Many of the nitrogen-containing specialty surfactants have anti–dye transfer properties along with other detergent properties. Using surfactants such as these will give the formulator a multifunctional product. Dye transfer data for commodity and specialty surfactants are shown in Table 6.

4. Synergistic Cleaning Performance

Again, the specialty surfactants are much too expensive to consider them the primary surfactant for prewash systems. For them to be useful in the marketplace, they need to work in conjunction with the anionic and nonionic detergent surfactants.

The specialty surfactants can be used as additives. They will enhance the

TABLE 5   Dye Transfer Test Method

Wash water temperature	50°C
Water hardness (wash and rinse cycles)	110 ppm (Ca/Mg)
Rinse water temperature	25°C
Wash time	40 min
Rinse time	3 min
Drying	Air dry

TABLE 6   Dye Transfer Inhibition for Surfactants

Surfactant	Avg. delta E
PVP	12.7
SLES	30.6
Na DBS	34.5
C12-15 alcohol ethoxylate (7)	18.0
NPE 9	16.7
Ethoxylated tallow amide	19.1
PEG-5 tallow amine	19.1
PEG-10 tallow amine	7.3
PEG-5 coco amine	8.9
Dihydroxyethyl tallow glycinate	7.1
Cocamidopropyl betaine	9.1
Lauramine oxide	6.6
Cocamidopropy hydroxysultaine	14.4

properties of the prewash system. Amides, amphoterics, and sulfosuccinates are used for viscosity control, foaming, wetting, and cleaning. Ethoxylated amines are useful for emulsification, dispersing, and dye transfer inhibition. By blending the specialty and standard surfactants together as shown in Table 7, a positive synergistic cleaning effect can occur.

5. Cost Performance Benefit

Due to the relatively high cost of specialty surfactants when compared to the typical cost of the anionic and nonionic detergents, a cost study should be done to obtain a balanced formulation picture. Many prespotters contain 20% solids, most of which is surfactant. Since most consumers greatly overdose a prespot stain treatment in an emotional desire to save their garment, low-solids formula-

TABLE 7 Synergistic Cleaning Performance

		% Dust-sebum removed	
Ratio	Surfactants	SLES	Blend
80:20	SLES: C12-15 alcohol ethoxylate (7)	58.0	60.2
65:35	SLES: PEG 10 tallow amine	48.2	55.7
80:20	SLES: coco diethanolamide (2:1)	62.7	65.9
80:20	SLES: sodium cocoampho propionate	47.7	49.9
		NaDBS	Blend
50:50	Na DBS: C12–15 alcohol ethoxylate (7)	49.5	55.0
80:20	Na DBS:coco diethanolamide (2:1)	50.9	53.9
65:35	Na DBS: PEG 10 tallow amine	27.1	37.8
		C12–15 alcohol ethoxylate (7)	Blend
65:35	C12–15 alcohol ethoxylate (7): PEG 10 tallow amine	59.0	62.8
50:50	C12–15 alcohol ethoxylate (7): PEG-30 glyceryl cocoate	60.5	63.6

tions containing specialty surfactants can be very cost-effective. The cost study should indicate that specialty surfactants can compete on performance with the anionic and nonionic surfactants.

Even though the specialty surfactants will never fully replace the conventional surfactants in prewash products, these products can be used cost-effectively.

VI. FORMULATIONS

Currently, the bottled prewash products on the store shelves do not vary much. The major producers have their products in a "trigger" spray, stain stick, or a container with a scrub brush on top. Aerosol products, once a familiar item on store shelves, are a rare sight. The chlorofluorocarbons (CFCs) which are the propellant found in aerosol products have been banned due to their effect on the ozone layer. Other propellants that may be on the market are flammable and hazardous.

When prewash products were first introduced into the market, many products contained ethoxylated alcohols. With a need for continual improvement, prewash products have moved from using ethoxylated alcohols, hydrocarbons, and soap to using enzymes and noncommodity surfactants to microemulsions.

A. Liquid Prespotter Formulations

The most popular type of pretreatment products is the liquid. Examples of starting formulations are shown in Tables 8 to 11. Enzymes can be used with the proper stabilizing agents. To obtain a cost-effective product quicker, an experimental design program may help. The mixing procedures are standard, with good agitation needed to ensure proper mixing.

B. Prewash Stain Stick

The next prewash product is the stick which is a solid surfactant product that can be placed in a "push-up" tube or, more commonly, in a "screw-up" tube. Stain sticks have grown in popularity with the consumer for a variety of reasons:

TABLE 8   Liquid Formulation #1

Ingredients	Weight %
Ethoxylated linear alcohol	4–8%
Cocamidopropyl betaine	3–5%
PEG-5 coco amine	4–6%
Dye, fragrance, preservative	0.1–0.3%
Water	to 100%

TABLE 9 Liquid Formulation #2

Ingredients	Weight %
SLES	8–12%
Coconut diethanolamide (1:1)	1–3%
Disodium cocoampho diacetate	3–5%
Dye, fragrance, preservative	0.1–0.3%
Water	to 100%

TABLE 10 Liquid Formulation #3

Ingredients	Weight %
PEG-10 tallow amine	10–15%
TEA LES	8–12%
d-limonene	8–12%
Butyl cellosolve [25]	1–3%
Dye, fragrance, preservative	0.1–0.3%
Water	to 100%

TABLE 11 Liquid Formulation #4

Ingredients	Weight %
Carboxylated alcohol ethoxylate	3–8%
Cocamidopropyl betaine	3–8%
Ethoxylated linear alcohol	2–5%
Propylene glycol	5–10 %
Enzymes	0.5–1.0%
Dye, fragrance, preservative	0.1–0.3%
Water	to 100%

Easy to use
Place more cleaning product on the stain
Convenient packaging
Easy to handle
Environmentally safer
Less chance for spillage or ingestion

The component that holds the stick together is sodium stearate. Other ingredients incorporated into the formulation are enzymes, solvents, surfactants, and water.

When choosing a surfactant for a stain stick, having the following characteristics in the surfactant will aid in making a better product:

Melting point—stability of product and ease of use
Water solubility—how well does the product get removed in washing

Sample formulations along with a blending procedure are listed in Tables 12 to 14.

TABLE 12 Stain Stick Formulation #1

Ingredients	Weight %
Sodium stearate	8.0
Ethoxylated tallow monoglyceride	67.0
Propylene glycol	10.0
Butyl Cellosolve [25]	10.0
Water	5.0
Dye, fragrance, preservative, enzyme	q.s.

TABLE 13 Stain Stick Formulation #2

Ingredients	Weight %
Sodium stearate	8.0
Ethoxylated tallow MEA amide	67.0
Propylene glycol	10.0
Glycol ether	10.0
Water	5.0
Dye, fragrance, preservative, enzyme	q.s.

TABLE 14 Stain Stick Formulation #3

Ingredients	Weight %
Sodium stearate	8.0
Sulfosuccinate	30.0
Ethoxylated tallow MEA amide	37.0
Propylene glycol	10.0
Solvent	10.0
Water	5.0
Dye, fragrance, preservative, enzyme	q.s.

1. Blending Procedure

Use a jacketed tank with a water circulating bath at 80°C. Weigh surfactants, propylene glycol, and water into jacketed tank. Begin agitation. Slowly add sodium stearate. Continue agitation after stearate has been added. Turn bath heater off but continue circulating the water. Add solvent and continue stirring. When thoroughly mixed, perfume, dye, enzymes, and preservative can be added. Pour into desired container and allow to cool. If after cooling the stick is not firm or is too firm, then adjustments need to be made to the quantity of sodium stearate and the surfactants.

C. Prespotter Gel

A more recent product is the gel. The container for this is a tube with a plastic brush on top. This product is viscous enough to remain on the stain without dripping; but fluid enough to be rubbed onto the fabric. To enhance the removal of the stain, a scrub brush is employed to mix the prespotter with the stain.

Samples of starting formulations are shown in Tables 15 to 17.

D. Physical Properties

Once the formulations have been prepared, physical properties and stability evaluation should be determined. The following traits should be characterized:

TABLE 15 Prespotter Gel Formulation #1

Ingredients	Weight %
C12–15 alcohol ethoxylate (7)	6–8%
PEG-5 coco amine	8–12%
Polymeric thickener	0.2–0.5%
Dye, fragrance, preservative, enzyme	0.1–0.3 %
Water	q.s. to 100%

TABLE 16 Prespotter Gel Formulation #2

Ingredients	Weight %
SLES	10–12%
Cocamidopropyl betaine	5–7%
Polymeric thickener	0.2–0.5%
Dye, fragrance, preservative, enzyme	0.1–0.3%
Water	q.s. to 100%

TABLE 17   Prespotter Gel Formulation #3

Ingredients	Weight %
SLES	10–12%
Disodium	6–8%
Ricinoleamido MEA	
Sulfosuccinate	
Polymeric thickener	0.2–0.5%
Dye, fragrance, preservative,	0.1–0.3%
enzyme	
Water	q.s. to 100%

Freeze-thaw stability: 3-day cycle
Various temperature stability: room temperature, 5°C, 40°C over a 1-month pe-
   riod
Viscosity
% Solids
Appearance/color
Density or specific gravity
pH—5% in 50/50 IPA/water
Softening point for stain sticks
Hardness test (needle penetration test [24]) for stain sticks

By monitoring the physical properties of the prewash products, the formulator
will be able to identify which formulations are acceptable.

VII. FUTURE OF PRESPOTTER PRODUCTS

With the future emphasis on using less energy and less water, the laundry detergent surfactants will need to be different from what is used today. These products
will have to be lower foaming, rinse easily in very little water, and clean just as
well.

When the detergent industry went from phosphate to nonphosphate products
over 30 years ago, a demand emerged for a prewash product. Now, as the detergent industry is on the verge of required product change again, the prewash product will still be a viable necessity.

REFERENCES

1.   Inform 8(1): 7–14, 1997.
2.   HA Hagen. In: New Horizons, an AOCS/CSMA Detergent Industry Conference (RT
     Coffey, ed.). Illinois: AOCS Press, 1995, pp 57–62.

3. Barrett. U.S. patent 3,741,902 to Purex Corporation (1973).
4. Diehl et al. U.S. patent 3,950,277 to Procter & Gamble (1976).
5. Claus et al. U.S. patent 4,225,471 to Chemed Corporation (1980).
6. Letton. U.S. patent 4,260,529 to Procter & Gamble (1981).
7. Brooks et al. U.S. patent 4,347,168 to Procter & Gamble (1982).
8. Gipp. U.S. patent 4,595,527 to S.C. Johnson & Son (1986).
9. Clark. U.S. patent 4,909,962 to Colgate-Palmolive (1990).
10. Broze et al. U.S. patent 5,035,826 to Colgate-Palmolive (1991).
11. Chemical Marketing Reporter 251(4):4, 1997.
12. Courtesy of S.C. Johnson Company, Racine, WI.
13. Manufactured by United States Testing Co., Hoboken, NJ.
14. Manufactured by Atlas Electronic, Chicago, IL.
15. ASTM method D4265-83: "Standard Guide for Evaluating Stain Removal Performance in Home Laundering."
16. ASTM method D3050-87: "Standard Method for Measuring Soil Removal from Artificially Soiled Fabrics."
17. ASTM method E313-73.
18. Brooks et al. U.S. patent 4,347,168 to Procter & Gamble (1982).
19. McRitchie, Smith. U.S. patent 4,321,165 to Procter & Gamble (1982).
20. Jakobi, Lohr. In: Detergents and Textile Washing. New York: VCH Publishers, 1987, p 139.
21. Barrett, Flynn. U.S. patent 3,953,353 to Purex Corporation (1976).
22. Diehl et al. U.S. patent 3,950,277 to Procter & Gamble (1976).
23. ASTM method D5548-94.
24. ASTM method D1321-86.
25. Trademark of Union Carbide Company, Inc.

8

Dry Cleaning Surfactants

AL DABESTANI   Consultant, Dublin, Ohio

I. INTRODUCTION

Dry cleaning can be defined in a general way as the cleaning of textiles in a substantially nonaqueous liquid medium. The process uses special equipment and organic solvents with detergent added. There is no clear history regarding the origin of dry cleaning technology. It is believed that dry cleaning originated in France about the middle of the 19th century. A member of a dyeing and cleaning firm in Paris is credited with the accidental discovery that spilled lamp oil removed soil from a table cloth [1–3]. Apparently that is the reason dry cleaning was sometimes referred to as "French cleaning."

The merits of dry cleaning had been recognized as early as 1862. Thomas Love in his book *The Dyer and Scourer* wrote, "It [Camphene] cleans all sorts of silk fabrics, when very dirty, in such manner that nobody would ever think they had been wetted; to clean them properly, they must be passed through one, two, or even three separate liquors. . . . It neither changes or alters color, but it takes the dirt, oil, and grease out of silks, cotton, and wool assisted by labor."

The term "dry cleaning" derived from the old term "dry solvents" used to describe liquids that do not wet textile fibers the same way that water does. Although dry cleaning solvents do not appear to wet the fabric like water, textile fibers are wetted even more rapidly by the substantially nonaqueous systems than by water due to the low surface tension of the solvents. The important distinction between dry and wet solvents is that water, and principally hydroscopic solvents like alcohols, glycols, and other hydroxylic-containing compounds, cause hydrophilic fibers to swell, which results in wrinkling and shrinkage of fibers.

Dimensional changes that fabrics undergo as they swell in water are transmitted throughout the textile structure and cause fabric damage. Thus, the main advantage of dry cleaning lies in its fiber-preserving and fiber-retaining properties. Another advantage of dry cleaning is that the cleaning process is done at low temperature. A comprehensive survey on theory and practice of dry cleaning has been given by Martin and Fulton [4].

This chapter primarily outlines the fundamental and applied principles of the nonaqueous cleaning process and the specialty surfactants used in this process.

II. DRY CLEANING PROCESSING TECHNOLOGY

A. Process

Cleaning processes in nonaqueous systems follow the same principles as in aqueous systems, particularly in removing insoluble soil. The process is carried out in a specifically designed, closed dry cleaning machine. A classical dry cleaning machine for chlorinated hydrocarbons is composed of four major units: cleaning unit, drying unit, recovery unit, and distillation unit. These units are mostly combined into one compact machine.

In the cleaning process the soil is removed by either "batch" or "bath" operation. These operations differ in solvent circulation, moisture content, detergent type, and length of wash time. The batch operation uses a fixed quantity of solvent for one operation. The solvent is not circulated during the wash cycle, while it is in the bath method. The bath operation is also known as the "charged" system. In the bath system, 1% to 4% concentrations of detergent and water are added into the wash tank.

During the 3- to 15-min cleaning cycle, textiles are immersed and tumbled in the solvent in a machine much like a front-loading washing machine. The load and the solution are confined in a perforated rotating drum that supplies mechanical energy to accelerate the removal of soils and stains. The cleaning solution possesses only a limited soil-bearing property, so it is necessary to remove the soil by mechanical means. The solvent is circulated from the base tank, to the filter, the bottom trap, and back to the base tank, and continuously filtered during the cleaning/wash cycle to remove insoluble soils.

In the drying system after the cleaning cycle, the solvent is removed from the load and drained into a holding tank. The extraction step removes most of the solvent from the garments before they are tumbled with heated air. This second stage serves as the vaporization phase by the recovery system. In the distillation system, the solvent is further cleaned by either continuous or batch distillation. The distilled solvent is returned to the machine base tank or other storage tank for reuse.

B. Solvents

The term "solvent" is generally applied to organic compounds used on an industrial scale to dissolve or suspend other materials. Basically any organic liquid can act as a solvent. Selection of the proper solvent or development of suitable blend is possible when the interaction between solvent and solute is understood. Edelstein documented the use of camphene to remove grease and oil from fabrics as early as the 1600s [5]. Camphene was replaced by benzene by the mid-1800s until the petroleum distillate became available and was used as a dry cleaning solvent [4]. Dry cleaning gained popularity in the United States by 1900 due to the availability of gasoline. Pioneers like W. J. Stoddard and Lloyd E. Jackson investigated other petroleum solvents to replace gasoline without the undesirable features of gasoline. Gasoline was replaced by Stoddard solvent by the early 1900s. Until 1928, low-flash-point and flammable solvents were the major solvents used in dry cleaning, and many establishments experienced fire and explosions caused by discharge of static electricity during the cleaning and drying process.

Nonflammable chlorinated hydrocarbon solvents were developed in 1930. Tetrachloromethane (carbon tetrachloride) was the first chlorinated solvent used for dry cleaning. It was introduced in Europe and gained popularity in the United States by the late 1950s. Trichloroethylene is less toxic and more stable than carbon tetrachloride, but causes more dye bleeding of many acetate dyes. These solvents were gradually replaced by perchloroethylene due to their toxic nature. Chlorinated hydrocarbons have better solvency than the corresponding nonchlorinated solvents. Increasing the number of chlorine substituents improves the solvency, but also increases the toxicity and reduces the combustibility and water solubility. Halogenated solvents have some tendency to hydrolyze in presence of water. Addition of organic bases as stabilizers prevents hydrolysis [6].

In 1960 a new generation of chlorofluorinated solvents, such as trichlorotrifluoroethane, was introduced to the dry cleaning market. They did not gain a significant market share, however. The most widely used solvent in today's dry cleaning is still tetrachloroethylene (perchloroethylene), or "perc."

There are a number of requirements that a dry cleaning solvent must meet in order to be acceptable [4,7]. An acceptable dry cleaning solvent has a high affinity for grease and oil and a low affinity for fabric dyes. It must be noncorrosive to metals commonly used in dry cleaning machinery as well as chemically and thermally stable under a variety of conditions. They must be nonflammable and sufficiently volatile to permit easy drying, and easily purified by distillation as well as relatively safe in terms of human health and toxicity. A summary of physical properties of selected dry cleaning solvents is shown in Table 1 [8].

As the safety of chlorinated solvents has become an issue in recent years, the idea of using liquid CO_2 as replacement for conventional dry cleaning solvents has flourished. High-pressure equipment is required for use of supercritical carbon dioxide. There are not sufficient data to support the efficiency of this solvent at this time, since developmental work has remained confidential.

TABLE 1 Comparison of Dry Cleaning and Solvent Properties

	1,1,1-Trichloro-ethane	Petroleum solvent	Perchloro-ethylene	Solvent-11	Solvent 113
Chemical formula	$CH_3\text{-}CCl_3$		$CCl_2=CCl_2$	CCl_3F	$CCl_2F\text{-}CClF_2$
Boiling point	165°F	300–400°F	250°F	23.7°C	47.6°C
Volatility	91	6	27	43.5	280
Latent heat of vaporization (BTU/lb)	102	122	90	1.49	63
Density g/cg³ (20°C)	1.32	77	1.62	48	1.58
Solvent power (KB)	101	26–45	90	none	31
Threshold limit value	350	100	25		1000
Flash point (min. temp.°F)	none	100–140	none	none	none
Water solubility (wt%)	0.07	0.01	0.015	0.0096	0.009
Surface tension (dynes/cm)	25.6	25–27.6	32.3	19	17.4–19.6
Use in dry cleaning (%)	<1	30–30	75–80	<1	1–2

C. Cleaning Aids

Cleaning aids are used to improve dry cleaning results. As in the normal home laundering process where a detergent is a combination of surfactants and suitable additives, a cleaning aid is a combination of several substances whose overall joint action produces the desired cleaning properties. Initially, no detergents were used in the dry cleaning process. Just before 1900, dry cleaners were searching for detergents which could enhance cleaning. There is a German abstract in 1893 Textile Colorist on "Improvement in Dry Cleaning," describing a mixture of olive oil, soap, benzene, and oil of terpene added to the bulk solvent to aid in stain removal.

Until about 1950, only two types of dry cleaning detergents were used—true soaps and filter soaps. True soaps were simply colloidal sols or gels composed of soaps and fatty-acid mixtures. Filter soaps were so called because they were soluble in solvent and consequently passed through the filter unit. Filter soaps were petroleum and other types of sulfonates. Today sulfonates of mixed petroleum hydrocarbons have been replaced by purer synthetic detergents.

Progress in dry cleaning detergent formulations has been much slower than in aqueous-base detergents perhaps due to the absence of reliable and standardized test methods. The best method may be to clean test swatches in a Launder-ometer with the chosen solvent and detergent system.

Detergency, particularly insoluble soil removal, in nonaqueous solvents follows the same principles as in aqueous systems. The differences are in the processes of water-soluble and oil-soluble soils removal. In the aqueous detergency process, the water-soluble soils are removed by water while the detergent helps remove the oil-soluble soils. In the nonaqueous process the reverse is true where the major function of detergent is the removal of water-soluble soil because the oil-soluble soil is removed by the solvent. The organic solvents frequently used in dry cleaning can dissolve a small quantity of water, and the addition of surfactants to the solvents enhances the solubility of water in the solvent. The removal of solid soil particles in dry cleaning is just as serious a problem as it is in the aqueous laundering process. The removal of particulates is promoted by use of detergents in both processes.

A summary of literature on detergency has been collected by Cutler and Davis. [9] A dry cleaning detergent must perform many functions in the dry cleaning process. The detergent must enhance the removal of insoluble soil and to prevent it from redeposition on the fabrics in the bath by dispersing or peptizing insoluble soils. It should keep the soil in suspension while it is being flushed out of the fabric and pumped through the filter.

Emulsification of water in the solvent is another function of the dry cleaning detergent to promote removal of water soluble soil by emulsified and solubilized water.

To meet these requirements a detergent should have the typical surfactant molecular structure with an appropriate hydrophilic-lipophilic balance (HLB). Almost any oil-soluble compound that is capable of forming a colloidal solution in a dry cleaning solvent is potentially a dry cleaning detergent. It requires that the detergent form water-in-oil micelles in solution and the micelles can solubilize water-soluble soils.

A formulated dry cleaning detergent or cleaning aid may contain the following components:

Primary surfactant
Cosurfactant
Solvents
Cosolvent
Organotropic (organic solubilizing) agent
Hydrotropic (water solubilizing) agent
Water and salts
Special additives

1. Solvents

Solvents are used in dry cleaning detergents to make them liquid and easy to dissolve in the bulk dry cleaning solvent.

2. Cosolvents

Addition of isopropylene, low-molecular-weight glycol ethers, cyclohexanol, certain amine compounds, fatty acids, and phosphate esters as cosolvents to a detergent blend is necessary to incorporate water in water-containing detergent blends. It is particularly important if water is added to the dry cleaning system with the detergent to facilitate the removal of water-soluble soils.

3. Organotropic Compounds

A wide range of compounds can be used, such as long-chain alcohols, ethers, and amines.

4. Hydrotropic Compounds

Sodium xylene sulfonate (SXS), short-chain alcohols, or glycol ethers help make the formulation homogeneous as well as solubilize water during the cleaning process. Ethylene glycol monobutyl ether can serve the function of hydrotrope, organotrope, and cosolvent.

5. Water and Salt

The water content of most cleaning aids range from 5% to 15%. Unfortunately, left over from their manufacturing process, many surface active substances contain a certain quantity of salts. The salt concentration should really be as small as possible. The less the salt content in the product, the more water-soluble soil can be dissolved during the cleaning phase.

6. Special Additives

Additives such as softening, retexturing (sizing), antistatic, oxidizing (bleaching), disinfecting, and optical brightening agents can be beneficial in cleaning aids. The importance of antistatic agents as antiredeposition agents in dry cleaning has been studied by Grunewalder and Muller. [10] Most commercial dry cleaning detergents contain one or two of the special additives.

III. OVERVIEW OF SURFACTANTS

A. General Classes

The most important and usually the most effective constituent of a cleaning aid is a surface-active substance. Historically, six classes of surfactants have been used in dry cleaning detergents [11]:

1.   Alkyl phenol and alkyl benzene sulfonates
2.   Sulfated fatty alcohols and sulfated oleic or ricinolieic acid

3. Petroleum sulfonates
4. Cetyl pyridinium bromide and other cationic agents
5. Esters of long-chain fatty acids with low- molecular-weight hydroxycarboxylic acids, such as stearyl tartrate
6. Oil-soluble nonionic agents, such as low HLB alcohol ethoxylates

As the list above indicates, very few oil-soluble surfactants fail to fit one of these classes.

Analysis of commercial dry cleaning detergents by the National Institute of Drycleaning (NID), now called the International Fabric Institute (IFI), indicates that a typical dry cleaning detergent contains the following types of surfactants [1,12–17]:

Ethoxylated alkylphenols
Ethoxylated phosphate esters
Ethoxylated alkanolamides
Sodium alkylbenzene sulfonates
Sulfosuccinates acid salts
Amine alkylbenzene sulfonates
Petroleum sulfonates
Soaps/fatty-acid mixtures
Fatty-acid esters of sorbitans

There are no comprehensive data available to judge the performance of these compounds used in dry cleaning detergents. Performance data and formulations are kept confidential by the involved companies, which is why there are so few references in the modern literature. Dry cleaning detergents are not highly formulated products and not very similar to home laundry detergents which contain a large portion of builder. Basically dry cleaning detergents are just surfactants or mixtures of surfactants. A commercial dry cleaning detergent may contain 40% to 90% active ingredients and be used at 0.5% to 4% in the dry cleaning solvent.

As stated, typical dry cleaning detergents contain surfactants, solvents, cosolvents, organotropic (dissolving) and hydrotropic (solubilizing) compounds, water, and salts.

Unlike aqueous systems, surface tension rarely poses a problem with cleaning solvents, but the right surfactants do help dissolve organic dirt with low solubility. In many systems, surfactants will form micelles.

Specialty surfactants provide an economical way to enhance and customize formulations. Most surfactants are suitable for traditional "perc" systems and the more current chlorofluoro systems. Detergent systems can also be used as prespotters on heavily soiled areas.

B. Alkanolamides

Alkanolamides based upon coconut, tallow, or oleyl radicals act as detergents, oil-in-water emulsifiers, lubricants, alkalinity builders (if excess alkanolamine is present), and conditioning agents. The amide function is known to cause slip and lubrication of equipment and to provide some cationic character for conditioning. Suitable alkanolamides include cocodiethanolamide and isopropanololeylamide. See Table 2 for a typical formula.

C. Alkylaryl Sulfonates

Alkylaryl sulfonates (ASs) have the properties of coupling agents, water-in-oil emulsifiers, fabric wetting agents, scouring agents, and degreasers. While LAs can work, amine salts of alkylbenzenesulfonates perform much better. See Table 3 for an example formula.

D. Alkyl Phenol Ethoxylates

Alkyl phenol ethoxylates are popular and act as oil-soluble detergents, oil-in-water emulsifiers, wetting agents, and scouring agents. Nonyl phenol ethoxylates

TABLE 2 Alkanolamide-Based Formula

Aromatic solvent	77%
Monoisopropanolamide of oleic acid	15%
Sodium 2-ethylhexylsulfosuccinate	8%

TABLE 3 Alkylaryl Sulfonate and Phosphate Ester-Based Formula

Mineral oil	50%
Isopropylamine alkylaryl sulfonate	30%
Phosphate ester of alcohol ethoxylate	10%
Water	8%
KOH (45%)	2%

TABLE 4 Alkyl Phenol Ethoxylate-Based Formula

Aromatic solvent	54.8%
Nonyl phenol ethoxylate (1–9 EO)	33%
Ethyleneglycol monobutyl ether	12%
Optical brightener	0.2%

with low amounts of EO are preferred, such as NPE-3. Table 4 has a typical formula. Possible replacements for NPE are low-HLB ethoxylated alcohols and ethoxylated amines.

E. Phosphate Esters

Phosphate esters are useful surfactants with the properties of coupling agents, oil-soluble detergents, oil-in-water emulsifiers, wetting agents, scouring agents, and lubricants. Due to the phosphorus atom they also function as antistatic agents and corrosion inhibitors. See Table 3 for a suitable formula.

F. Quaternaries

Quaternaries are unique surfactants and act as antistatic agents, detergents, oil-in-water emulsifiers, corrosion inhibitors, and lubricants. Cationic surfactants are used less often, but if used according to manufacturer's recommendations, provide rapid soil release and excellent water-soluble soil release [3]. They can be either oil soluble, water soluble, or dispersible depending on the molecular weight and presence of any fatty chains. While most commercial cationics contain one or more long alkyl chains, the most common types in dry cleaning are based upon diethylamine plus several moles of propylene oxide then quaternized with methyl chloride. Quats' main advantage is their ability to reduce static, which is even more important if a flammable solvent is used. See Table 5 for one possible formula.

G. Sulfosuccinates

Sulfosuccinates have detergent properties and act as oil-in-water emulsifiers, solubilizers, coupling agents, and excellent wetting agents. Tables 2 and 6 show starting formulas.

IV. DETERGENCY IN NONAQUEOUS MEDIA

The dry cleaning solvent removes oily soils easily but the addition of detergents greatly enhances the removal of other soils present on the substrate. The solvent

TABLE 5 Quaternaries-Based Formula

Kerosene	67%
Polypropoxy quaternary ammonium chloride	13%
Alcoholethoxylate (5 mole EO)	9%
Oleic acid	3%
Water	3%

TABLE 6   Sulfosuccinate-Based Formula

Aromatic solvent	60%
Sodium dioctyl sulfosuccinate	14%
Nonyl phenol ethoxylate	10%
Ethylene glycol mono butyl ether	7%
Amine salt of alkylarylsulfonate	6%
Isopropyl alcohol	2%

dissolves the oily soils and releases the insoluble or pigmentary soils. These can be redeposited on the substrate unless they are forced to remain in suspension [18]. The mechanism of redeposition of insoluble soils is due to high interfacial energy between most fibers and the dry cleaning solution. Reduction of interfacial energy occurs by adsorption of surfactants at the substrate-solution interface. Mechanisms have been postulated based on zeta potential and other surface electrical phenomenon [4].

Reduction of surface tension generally does not apply well to oil-soluble surfactants in a solvent system. However, the phenomena of micelle formation and solubilization have been observed and are important in nonaqueous systems [19]. Most of the published literature on micelle formation and solubilization concerns aqueous systems. Unfortunately, the knowledge and literature on the critical micelle concentration (CMC) in nonaqueous micelle systems are still limited.

A.   Micelle Formation in Nonaqueous Solutions

Unlike surface tension reduction, the phenomenon of micelle formation in nonaqueous systems is significant as this leads to the phenomenon of solubilization [20]. Evidence of micelle formation in nonaqueous solutions of detergents was reported by Arkin and Singleterry [21]. Their work was based on a technique that certain dyes fluoresce about the CMC [22]. Arkin and Singleterry showed that with a solution of dye and soap in benzene illuminated by a plane of polarized green light, emitted light was 28% polarized. A dye-and-soap solution of equal concentration in methanol exhibited only 2.3% polarization. They concluded that the difference was due to dye molecules being absorbed on the soap micelles so that the apparent volume of the dye molecules is increased. The fluorescence increases proportionally and can be calculated by Perrin's relation [23]. X-ray evidence of potassium oleate micelles in benzene supported the ideas that soap and alcohol form mixed micelles in benzene [24]. Another study of micelle formation by sulfonates in dry cleaning hydrocarbons was done by viscosity measurements [25].

Formation of micelles in nonaqueous solvent systems is due to interaction of numerous intermolecular forces. A surfactant molecule consists of a hydrophilic polar moiety and an oleophilic tail. In a nonpolar medium the head group of the surfactants are oriented inward toward the water while the nonpolar tail is in the solvent [26]. These are usually referred to as "inverse micelles."

Micelles of detergents are organized groups of oriented molecules that form spontaneously. They can be either spherical or lamellar shape of various sizes, depending on the ionic nature and the concentration of the surfactant [27]. These inverse micelles exhibit unusual properties; the most useful is that the hydrophilic interior of these micelles is capable of dissolving water [19]. Solubilization [28] of water in a solvent system having micelles with the hydrocarbon moieties concentrated at the exterior of the micelles has been observed.

Surfactant molecules in a dry cleaning "charge system" can form colloidal aggregates at or above their CMC.

B. Factors Influencing Cleaning Performance

Many factors other than the amount and the nature of detergent and cleaning aids influence the cleaning efficacy in the dry cleaning process; the water content of the bath, the relative vapor pressure of the water in the bath, and the electrical conductivity of the bath are worth mentioning. Studies at IFI indicate that if salt is used as a criterion, the use of stronger charge in both petroleum and synthetic solvent is beneficial [29,30]. The results of these experiments are shown graphically in Figure 1.

Solvent temperature is another factor that must be controlled when using detergents. There are several reasons for this. The vapor pressure of water in the solvent increases sharply with temperature. This effect is illustrated in Figure 2 [31]. Another effect of temperature is the decrease of micelle formation in colloidal solutions, which reduces the detergency.

C. Vapor Pressure of Water in Solution

It is important to lower the vapor pressure of water in the dry cleaning process so that absorption of water by the fabric is limited to the same amount that would be absorbed in air at the same relative humidity. This eliminates shrinkage and other types of water damage. Solvents interact with fabrics differently from water. Many fibers, especially hydrophilic fibers, such as wool and rayon, absorb water. As water wets these fibers, the textiles suffer undesirable dimensional changes.

The vapor pressure of water in the solvent is temperature dependent, so solvent temperature control is required. The water solubility of water-soluble substances in the dry cleaning bath depends on both the absolute amount of water present in the cleaning solution and the partial pressure of the water in the vapor phase in

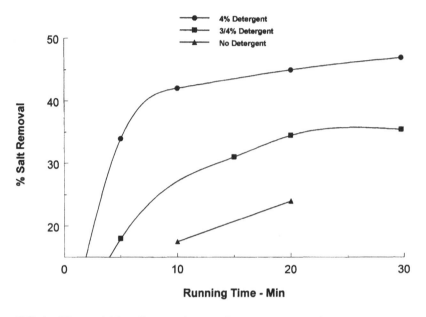

FIG. 1 Water-soluble soil removal versus detergent concentration.

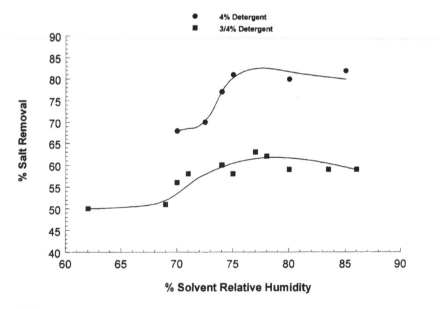

FIG. 2 Water-soluble soil removal at different solvent relative humidities.

equilibrium with the solution. In the dry cleaning process the relative partial pressure of water in the solvent is measured by relative humidity in the solvent. Absorption of water by textiles has been studied and reported in literature [32,33].

A solution of detergent in a hydrocarbon containing solubilized water is a three-component system of two phases. The term "solvent relative humidity" has been discussed by Martin and Fulton [4]. Solubilization of water in dry cleaning solvents has been studied by several researchers [34–36].

D. Mechanism of Soil Removal and Redeposition in Dry Cleaning

The soils encountered on textile fabric may be classified according to their solubilities which ultimately determine their best way of removal.

1. Removal of Oil-Soluble Soils

Some soils, such as oils and greases, are solvent-soluble soils which are insoluble in water but soluble in certain organic solvents. These types of soil in home aqueous detergent systems are removed by emulsification of the soil. Their removal in nonaqueous systems is by the solvent. There are insufficient data regarding the exact mechanism of solvent-soluble soil removal by detergents and solvents. The cleaning efficacy of solvents is known as relative solvent power that can be determined by an empirical rating system called Kauri-butanol (KB) test [37]. As indicated in Table 1, chlorinated solvents have higher solvent power than either hydrocarbon or chlorofluoro solvents.

2. Removal of Water-Soluble Soils

These types of soils range from completely soluble simple salts and sugars to the more sparingly soluble starches and albumins. Their removal in an aqueous system is by water, but in a nonaqueous system removal is accomplished by a combination of solvent, detergent, and controlled amount of moisture by either solubilization or emulsification. The removal of polar soils, such as sodium chloride and glucose, is a function of solvent relative humidity [29]. Experiments (Fig. 1) have shown that increase in detergent concentration up to 4% as used in charge system removes more water-soluble soil [30]. The efficiency of water-soluble soil removal in a charged system is the effective reduction of vapor pressure of solubilized water by detergent micelles which prevent the removal of water by absorption on the garment.

Studies by Martin and Fulton [4] appear to substantiate the use of higher detergent concentration in both petroleum and synthetic solvent if salt removal can be used as criterion. They cited that the removal of sodium chloride or glucose, measured as function of relative humidity, sharply increases in the region of 75% solvent relative humidity (SRH) at 25°C. In Figure 2 the removal of salt as a function of solvent relative humidity is shown. The sharp increase in the vicinity of 75% SRH is quite evident. Figure 3 shows that higher detergent levels allow a higher water content in the dry cleaning system without getting the SRH in the "danger zone" where clothes shrink.

3. Removal of Insoluble Soils

The third class of soils comprises materials that are soluble in neither aqueous nor nonaqueous systems. In both aqueous and nonaqueous systems, these materials are removed by peptization. The amount of literature on either removal or redeposition of insoluble soils in nonaqueous systems is scanty. In dry cleaning, the moisture content of fibers is the prime factor in soil redeposition, as the moisture content may vary from almost completely dry to up to the amount of regain moisture that they normally contain at a relative humidity of 75%.

Piment soils consist of solid particles of various sizes. They are generally insoluble in dry cleaning systems. In most cases, carbon black is the principal member of this group. Wendell experimented with cationic, nonionic, and anionic surfactants with different HLBs to determine the influence on the carbon black removal and suspension in solvents system [38]. He concluded that the nature of suspending agents, concentration, and water content of the system highly influence carbon

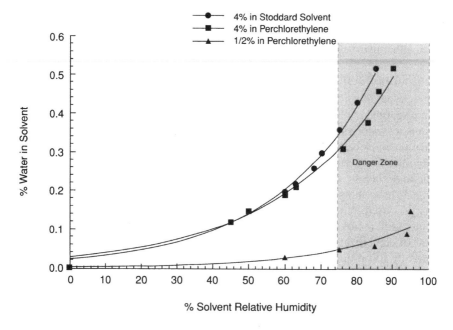

FIG. 3 Effect of detergent level on water content and SRH.

black removal from fabrics. He also found that anionic surfactants with a carbon chain length of 16 to 18 were most potent in removing this type of soil [39]. The two performance tests found most reliable in NID are graying (soil redeposition) and removal of water-soluble soil. Cleaning and graying are determined with a reflectometer and compared to the reference. For a composite soil made of pigment and oil, the oily part is removed by solvent and most of the pigment particles float off. The pigment particles that adhered to fabric after removal of oil cannot float off. The fiber type is important for pigment soil removal. Pigment particles are removed from wool and cellulose fibers more easily than from synthetic fibers.

V. CONCLUSION

As the formulas in Tables 2 through 6 illustrate, the specialty surfactants used in dry cleaning vary widely and almost anything will work to some extent. Anionics, cationics, nonionics, and blends all function. Anionic-cationic blends should be avoided for the obvious reason of forming insoluble complexes.

As mentioned previously, the literature on nonaqueous cleaning process is inadequate. More research is required to better understand the cleaning mechanism in solvent systems, particularly if liquid CO_2 systems become commercial.

REFERENCES

1. EM Michelson. Remember the Years: 1907–1957. National Institute of Dry Cleaning, 1957.
2. CF Goehring. Textile colorist. Chemiker-Zietung, May 1894.
3. Focus, International Fabric Institute, 14(4), November 1990.
4. AR Martin, GP Fulton. Dry Cleaning Theory and Technology. New York: Wiley, 1958.
5. SM Edelstein. American Dyestuff Rep 46:1, 1957.
6. K Johnson. Dry Cleaning and Degreasing Chemicals and Processes. Park Ridge, NJ: Noyes Data Corporation, 1973.
7. Commercial Standards CS3-41, National Bureau of Standards, U.S. Department of Commerce, Washington, DC, 1928.
8. Ansi/ASTM D 484-76, Ann. Book of Standards, part 29 (1977).
9. WG Cutler, RC, Davis. Detergency Theory and Test Methods. New York: Marcel Dekker, 1972.
10. W Grunewalder, H Muller. Faeber-Zeitung, 1965.
11. Schwartz, Perry. Surface Active Agents. New York: Wiley Interscience, 1949, p 464.
12. WC Powe, WL Marple. J Am Oil Chem Soc 37:136, 1960.
13. KH Bey. Am Perfume 79(8):35, 1960.
14. WC Powe. Text Res J 29:879, 1959.
15. Commercial Standard CS3-41, National Bureau of Standards, U.S. Department of Commerce, Washington, DC, 1928.

16. CB Brown. Research 1:46, 1947.
17. W Brown. Modern Dyeing and Cleaning Practice. London: Heywood & Co., 1937.
18. J Berch, H Peper, GL Drake. Text. Res. J. 34:29, 1964.
19. CM Aebi, JR Wienbush. J Colloidal Sci 14:161, 1959.
20. McBain, Merrill, Vinograd. J Am Chem Soc 62:2880, 1946.
21. Arkin, Singleterry. J Am Chem Soc 70:3965, 1948.
22. Corrin, Harkins. J Am Chem Soc 69:679, 1947.
23. Perrin. J Physique 7:390, 1926.
24. Schulman, Riley. J Colloid Sci 3:383, 1948.
25. Van der Waarden. J Colloid Sci 5:448, 1950.
26. FM Fowkes. Solvent Properties of Surfactants. New York: Marcel Dekker, 1967, p 67.
27. K Kon-no, A Kitahara. J Colloidal Interface Sci p 636, 1971.
28. JW McBain. Adv Colloid Sci 1:99, 1942.
29. NID Technical Bulletin T-292, Adco Corp., October, p 172 1952.
30. NID Technical Bulletin T-329, Adco Corp., June, p 173, 1954.
31. NID Technical Bulletin T-329, Adco Corp., June, p 175 1953.
32. AR Martin, P Buhl. J. Colloid Interface Sci 29:42, 1969.
33. M Wentz, DM Cates. Text Res J 48:166, 1978.
34. MB Mathews, E Hirschhorn. J Colloidal Sci 430, 1959.
35. GP Fulton, JC Alexander, AC Lloyd, M Schwartz. ASTM Bulletin, 1953.
36. M Wentz, WH Smith, AR Martin. J Colloidal Interface Sci, 1969.
37. Book of ASTM Standards, Part 29, D-1133 (reapproved 1973).
38. H Wendell. Melliand, 1959–1960.
39. M Wentz. Detergency Theory and Technology. New York: Marcel Dekker, 1987, p 478.

9

Concentrated and Efficient Fabric Softeners

AMJAD FAROOQ, AMMANUEL MEHRETEAB, and JEFFREY J. MASTRULL   Colgate-Palmolive Company, Piscataway, New Jersey

RÉGIS CÉSAR and GUY BROZE   Colgate-Palmolive Research and Development, Inc., Milmort, Belgium

I. INTRODUCTION

Detergents are designed to remove the soils and stains from whatever substrates they clean be it clothes, hair, or a hard surface. For example, the laundering of clothes leaves them clean, but with a harsh feel and a slightly elevated pH. This phenomenon is exaggerated further when high-caustic industrial and institutional (I&I) detergents are used. The washed garments are also prone to collect static electricity, particularly if dried in an electric tumble dryer. These side effects of the cleaning process are very undesirable to the consumer. Fabric softeners are specialty surfactants used to improve the fabric feel or handling and to impart a pleasant smell to the laundry after washing [1]. Household fabric softeners were first introduced on the U.S. market in the mid- to late 1950s and were launched in Europe and Japan starting in the 1960s by detergent manufacturers.

Similar undesirable side effects result with the shampooing of hair and have resulted in the development of hair conditioners. This chapter will deal with fabric softeners, although the organic and physical chemistry is similar for hair conditioners.

II. FABRIC SOFTENER TECHNOLOGY

There are three different ways of treating clothes with a fabric softener in the laundry process: rinse cycle; wash cycle; dryer cycle.

Rinse cycle softeners are the most common and efficient where the fabric softener surfactant is added after the wash cycle at the beginning of the rinse cycle of a washing machine. Fabric softener molecules can also be impregnated on thin cloth sheets and placed in with the clothes when drying by an electric tumble dryer. Fabric softeners have been formulated into both powdered and liquid detergents for one-step treatment of clothes in the laundry [2–4]. These are so-called detergent-softeners, or "softergents." They are very convenient and were popular in the early 1980s.

The most common active ingredients in fabric softeners are cationic surfactants. Traditional examples are dihydrogenated tallow dimethyl ammonium chloride (DHTDMAC), amido-imidazoline quaternaries, and difatty diamidoamine quaternaries like bis(alkyl amidoethyl)-2-hydroxyethyl methyl ammonium methyl sulfate (see Fig. 1).

Because fabric softener surfactants are cationic they are generally incompatible with traditional anionics used in detergents. Thus the softener is normally applied in the rinse or drying portion of the laundry process. When a workable detergent-softener is prepared, it requires anionic free formulations using specialty surfactants. Ethoxylated alcohols are compatible with cationics and perform reasonably well as the detergent in a detergent-softener formulation. Nitrogen-containing specialty surfactants like ethoxylated amines, ethoxylated amides, amine oxides, and amphoterics/betaines perform somewhat better in cleaning and softening [2]. In most cases, however, both detergency and softening performance is less than premium, so despite the convenience, consumer popularity for detergent-softeners has declined. This chapter will focus on new developments in rinse cycle fabric softeners.

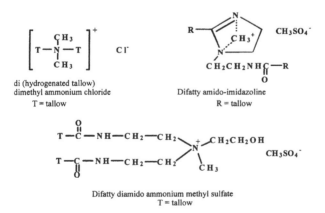

FIG. 1 Traditional fabric softener molecules.

Within the last 5 to 8 years, there has developed a need for efficient fabric softening agents with better environmental compatability than the most extensively used softening material, DHTDMAC [1,5]. Several new types of fabric softener molecules which exhibit rapid biodegradation were developed [6,7]. The esterquat shown in Figure 1 is based upon triethanolamine (TEA), while similar diesterquat molecules are in use based upon methyldiethanolamine (MDEA) and 1-dimethylamino-2,3-propanediol (DMAPD).

In general, all fabric softener molecules possess a positively charged nitrogen atom and two long-chain alkyl hydrophobes. The two long chains provide for the softening and "slick" feel, but result in very low true water solubility, on the order of 1 ppm. These surfactants are water dispersible, however. For three decades, rinse cycle fabric softeners were essentially 3 to 8 wt% active dispersions [8], which was considered regular strength. Traditionally, large consumer products companies used DHTDMAC while private-label and smaller companies formulated with the imidazoline and amidoamines quaternaries. Consumer fabric softeners were formulated at neutral to slightly acidic pH (5.5 to 7.5). The I&I industry uses all the types of softener molecules, but formulates "softener sours" at low pHs. Softener sours contain acids like phosphoric to give a softener dispersion that neutralizes the residual basicity of I&I detergents.

Several factors can affect the softening of a particular dispersion:

Chemical structure of the softener molecule (melting point, molecular weight, steric hindrance, dispersbility)
Water temperature during the rinse cycle
Fabric softener concentration
Dispersibility by additives/mixing
Dryer versus air dry

Regarding softening of the inherent molecular structure, DHTDMAC > imidazoline quat > amidoamine quat ≥ TEA esterquat. As nature would have it, the ease of dispersibility is an opposite trend, where amidoamine quat > TEA esterquat > imidazoline quat > DHTDMAC. The soft tallow amidoamine quat disperses fairly easily in cold water while hard tallow DHTDMAC needs good agitation and hot water. Warm rinse water actually gives the best softening with a hard-tallow-derived molecule softening better than a soft-tallow analog. In cold water a more liquid quat is needed; thus a soft tallow derivative may be better than hard tallow if the hard version does not disperse easily. Since colder wash and rinse temperatures are the trend globally, dispersibility is even more important formulation criteria.

In the past 10 to 15 years, a market for less bulky, more concentrated products, "ultras," has emerged as a result of desire to cut packaging and transportation costs, reduce the shelf space, and utilize smaller, easier-to-handle containers. It is difficult, however, to prepare concentrated aqueous dispersions with more than about 20% of cationic softeners without encountering severe product viscosity and storage stability problems [9,10]. With the increase in concentration of the active ingredients of fabric softeners, the viscosity increases, which inevitably results in nonpourable products. Furthermore, the physical stability of the formulation decreases because of the increased amount of the dispersed material and dissolved electrolyte. The product efficacy and the dispersibility are also affected by the increased concentration. This is true for dihydrogenated tallow dimethyl ammonium chloride, amido-imidazolinium compounds, and difatty diamido ammonium methyl sulfate [9–11]. The approximate upper active level (wt%) limits for these softeners are 12, 20, and 26 (iodine value = 40), respectively, for a typical soft-tallow-based dialkyl softener.

For most dispersions, a dilute (1% to 8%) fabric softener will give the best softening per gram while an ultra (12% to 28%) will give worse softening unless special care is taken to make formulation dispersible. A number of additives have been observed to improve dispersibility:

Monoalkyl quats
Ethoxylated alcohols
Ethoxylated amines
Solvents

The monoalkyl quats can be alkyltrimethyl, alkyl methyl diethoxy, or a monoalkylesterquat. Since these are also quats, softening is not compromised much by their addition. Ethoxylated alcohols certainly improve dispersibility and reduce particle size, but too much hurts softening since it acts like a detergent and prevents deposition of the quat on the fabric. Proper selection and level of solvents like ethanol, isopropanol, propylene glycol, or hexylene glycol can improve the dispersibility of a fabric softener concentrate.

Eventually all dispersions or emulsions separate or get thick although this may take from 1 day to 5 years. Esterquats are inherently less stable molecules and hydrolyze, causing the viscosity to climb. Formulation pH is critical and best at 2 to 4 for esterquats. For a viable commercial fabric softener product, usually 6 months' shelf stability is needed.

The remainder of this chapter addresses the needs of the time and deals with recent developments by the authors in the preparation and properties of stable, and pourable, softener emulsions with solid loadings of 20% or more. This was done by employing a nonconventional softening material, difatty amido amine or bis-(alkyl amidoethyl)-2-polyethoxy amine (AA(H) and AA(S)), combined with difatty amido-imidazolinium methyl sulfate (AIm) and triethanolamine diester quaternary methyl sulfates [bis-(alkylcarboxyethyl)-2-hydroxyethyl methyl ammonium methyl sulfate] (DEQ and

DEQ (H)). Amidoamine (and diesterquat) in which the tallow group is hydrogenated is designated by an abbreviation "H." The soft-tallow-based amidoamine is referred by the symbol AA(S). DEQ is a custom-made diesterquat, which contains about 75% of soft-tallow moieties and 25% of hydrogenated tallow groups.

While a small amount of a monoalkylquat is known to help disperse a dialkylquat, this study focused on expanding that concept by combining an acid salt of a dialkylamidoamine with a dialkyl quat to synergistically disperse each other. Because both components are dialkyl, the goal was premium softening. (Figure 2 presents the structures of amidoamine and esterquat.)

AA(H); T = Hydrogenated Tallow or Varisoft-510)
AA(S); (T = Soft Tallow, Varisoft 511 or Varisoft 512)
x = 1.7-2.9

DEQ (H), and DEQ

Triethanolamine Diester Quaternary Methyl Sulfate

T=Soft Tallow

AIm

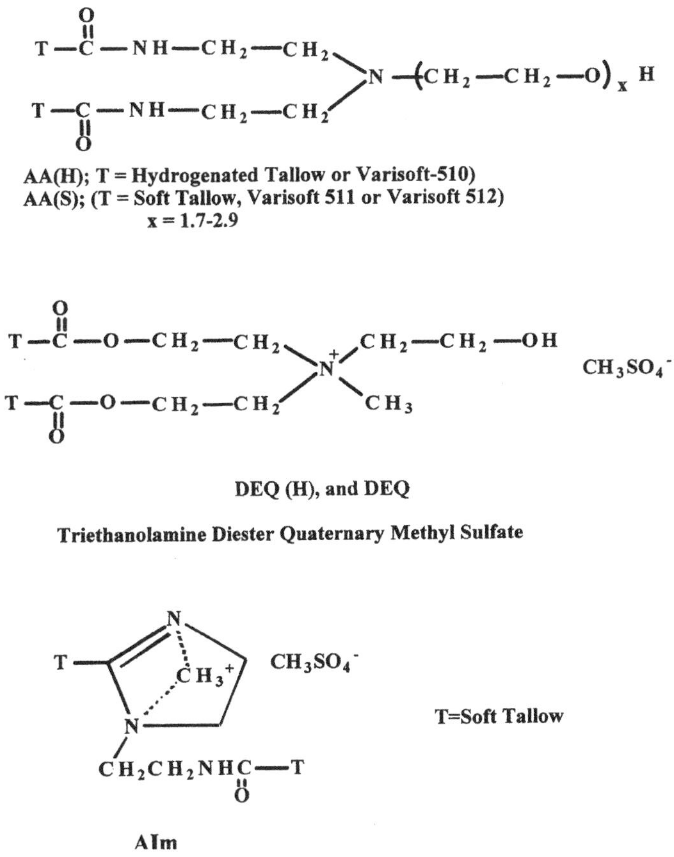

FIG. 2 Amidoamine and esterquat structures.

III. EXPERIMENTAL

A. Materials

Samples of difatty amidoamines (Varisoft 510, AA(H), 100% A.I.; Varisoft 511, AA(S), 100% A.I.; and Varisoft 512, AA(S), 90% active in isopropanol) were obtained from Goldschmidt Chemical Corporation. The AIm used was 90% active 4,5-dihydro-1-methyl-2-nortallow alkyl-(2-tallow amidoethyl) imidazolinium methyl sulfate by Akzo-Nobel. Quaternized triethanolamine diester methyl sulfate (DEQ; Tetranyl AT-75; 85%) sample was obtained from Kao Corporation Japan. DEQ (H) (100% hydrogenated diesterquat) sample was obtained from High Point Chemical Corporation. Armeen DMHTD, H-tallow dimethyl amine was obtained from Akzo Chemical Company. BP-7050 polymer was obtained from BP Chemical Limited, United Kingdom. This polymer is acrylamide/quaternary ammonium acrylate (50%) in solvent refined mineral base oil.

B. Methods

1. Rheological and Structural Properties

Viscosity was measured at 22 to 26°C with a Brookfield RVTD or Brookfield DVII viscometer in all measurements unless otherwise stated. Rheological measurements were performed on a Carri-Med CSL, Controlled Stress Rheometer at room temperature with parallel plate geometry (plate diameter, 4 cm, acrylic), utilizing Oscillation and Flow Packages. Average particle size of the dispersions was measured by LS Coulter 130 (suitable for 1 to 100 μm) or Brookhaven Laser Light Scattering Instrument (suitable for sizes < 2 μm) at 90° angle. The structure of softener dispersions was studied by CRYO scanning electron microscopy [Scanning Electron Microscope, Cambridge model S-360 equiped with a cold stage (Oxford Instrument model CT 1500C)] using the freeze-fracture replica technique.

2. Softening Test Methodology

The softening tests were performed in 350 ppm hard water (typical of European washing water) unless otherwise mentioned employing a laboratory mini softening machine (a device simulating the rinse cycle of a washing machine). The hard water was prepared by dissolving 7.5 g $CaCO_3$ in 22 L deionized water by saturating the system with CO_2 gas (the procedure involves the bubbling of CO_2 gas through the solution for 15 to 20 min and then thorough mixing by a magnetic stirrer until the cloudy system turns to a clear solution; usually this step takes about 3 to 4 hours). The standard conditions employed for the softening test were as follows: 4.4 g softening composition (5% by weight standard DHTDMAC dispersion or 3.5% to 5% by weight of AA or AA/AIm or AA/DEQ system) and 120.0 g cotton/L water, 5 min rinse, and line-drying overnight in a conditioned room (22°C, 50% relative humidity). The softness of the treated cotton terry towels was evaluated by trained judges using a method whereby the judges rated the softness from 1 to 10, with a rating of 1 being the harshest (untreated control) and 10 being the softest. Fabrics treated with standard material DHTDMAC were included as reference samples. The softening efficacy of the new softener dispersions (1:1 or concentrates) is reported by a scale EQ. EQ stands for equivalent amounts of DHTDMAC. The EQ is the DHTDMAC concentration of a hypothetical composition needed to deliver the same softness as the tested composition under the standard test conditions. The softening corresponding to 5% level of 100% active DHTDMAC is referred to as 5 EQ. If an ultra dispersion with 28% of softener actives diluted to 3.5% actives provides a softening equivalent to 5% of DHTDMAC, then the ultra dispersion exhibits a softening efficacy of 40 EQ.

C. Preparation of Dispersions

1. Preparation of 5% Dispersion of DHTDMAC

Melted 70°C oil phase (33.33 g) is slowly added to the hot water (466.67 g) at 70°C with stirring. Then NaCl solution (0.4 g of 10% NaCl solution; active concentration of NaCl in the emulsion, 0.008% or 1.368×10^{-3} M) is added and the stirring emulsion is cooled to room temperature with an ice/water bath. *Characteristics:* Brookfield Viscosity (spindle 4, 20 rpm, RT) = 40 cP; average particle size = 23 μm (LS Coulter 130); pH = 5.5.

2. Preparation of Difatty AA Emulsions (AA) and AA/Alm Emulsions

Compositions 1–17 (Tables 1–3) were prepared by adding the aqueous solution of HCl at 70°C into the molten oily phase (having propanol also) while stirring.

TABLE 1 Compositions of Varisoft-510, AA(H) Emulsions at 10% to 15% Levels in the Presence of H-Tallow Dimethyl Amine Emulsifier (Armeen DMHTD)*

	Composition			
	1 (wt%)	2 (wt%)	3 (wt%)	4 (wt%)
Varisoft-510, AA (H)	10	11	12	15
Armeen DMHTD	1.5	1.65	1.8	2.25
HCl	0.72	0.79	0.86	0.72
2-Propanol	4.0	4.4	4.0	—
CaCl$_2$	—	—	—	0.35
H$_2$O	balance to 100	balance to 100	balance to 100	balance to 100

*These data correspond to Figure 3.

TABLE 2 Compositions of Emulsions of AA(H)/AA(S) (16.4% to 28.4%)

	Composition								
	5 (wt%)	6 (wt%)	7 (wt%)	8 (wt%)	9 (wt%)	10 (wt%)	11 (wt%)	12 (wt%)	13 (wt%)
Varisoft-510, AA (H)	15	16	17	18	19	20	22	24	26
Varisoft-512, AA (S)	1.36	1.45	1.54	1.632	1.722	1.813	1.99	2.176	2.36
HCl	1.08	1.46	0.72	0.72	0.9	0.88	—	1.6	1.46
2-Propanol	4	4	4	4	4	4	4	4	4
CaCl$_2$	0.5	0.46	0.37	0.3	0.45	1.18	0.27	0.2	0.169
H$_2$O	balance to 100	balance to 100	balance to 100	balance to 100	balance to 100	balance to 100	balance to 100	balance to 100	balance to 100

TABLE 3 Compositions and Physical Properties of Emulsions of AA(H)/AA(S) and AA(H)/AIm

	Composition			
	14 (wt%)	15 (wt%)	16 (wt%)	17 (wt%)
Varisoft-510, AA(H)	10	10	12.5	
Varisoft-512, AA(S)	—	10		
AIm	10	—	12.5	20
HCl	0.72	1.08	0.9	
2-Propanol	4	4	5	
CaCl$_2$	0.37	0.38	0.45	0.27
H$_2$O	balance to 100	balance to 100	balance to 100	balance to 100
Physical properties				
Average particle size (μm)	1.1		1.1	11.7
Softening efficacy (EQ)	>24	<<20	35	<8
Viscosity (cP) 20 rpm, spindle 4-initial	60	30	140	120
Viscosity (cP) week 1	120	40	360	420

Stirring was achieved by a Premier mixer which was hooked to a Variac for control of the motor speed. The $CaCl_2$ solution was added into the hot dispersions which were subsequently allowed to cool to room temperature while stirring. During cooling, a few drops of Dow Corning 1430 antifoam was added.

3. Preparation of AA (S)/DEQ Emulsions

The aqueous dispersion of compositions 18–23 (Table 4) were prepared by adding a molten mixture of the DEQ and Varisoft 512, AA(S) at 35 to 70°C to an aqueous (deionized water) solution of HCl (and emulsifier for 23) at 60 to 70°C while stirring with a Premier mixer connected to a Variac for control of the motor speed. $CaCl_2$ solution was added into the hot dispersions, which were allowed to cool to room temperature slowly with stirring. During cooling, a few drops of Dow Corning 1430 antifoam was added.

4. Preparation of AA(H)/DEQ Emulsions

Compositions 24–26 (Table 5) were prepared by initial mixing at 1000 to 1100 rpm (employing a 3-pitched blade, propeller stirrer, diameter of stirrer, 70 mm; diameter of shaft 8 mm, length 350 mm; Tekmar RW 20 DZM mixer) the oil

TABLE 4 Compositions and Physical Properties of Concentrated Emulsions of Varisoft-512 (AA(S)) and (DEQ)

	Composition					
	18 (wt%)	19 (wt%)	20 (wt%)	21 (wt%)	22 (wt%)	23 (wt%)
Varisoft-512	12	12	15	16	17	17
DEQ	12	12	15	16	17	17
Synperonic A20	—	—	—	—	—	2
HCl	0.87	0.87	1.08	1.08	1.22	1.22
BP-7050	—	0.02	—	—	—	—
$CaCl_2$	0.78	0.4	1.25	0.8	1.09	0.95
Dow 1430 Defoamer	0.03	0.03	0.02	0.04	0.04	0.04
H_2O	balance to 100	balance to 100	balance to 100	balance to 100	balance to 100	balance to 100
Physical properties						
Softening efficacy EQ	—	24	—	—	~35	~32
Average particle size (μm)	—	5.9	—	—	1.7	2.3
Viscosity (cP) 20 rpm, spindle 4-initial	90	250	850	1190	820	1370
Week 10	140	250	640	—	430	1040

TABLE 5 Compositions and Physical Properties of Concentrated Emulsions of AA and DEQ

	Composition			
	24 (wt%)	25 (wt%)	26 (wt%)	27 (wt%)
Varisoft-510, AA(H)	17	17	17	13.1
Varisoft-512, AA(S)				4.38
DEQ	17	17	17	10.5
Douscent 653				2.4
Preservative				0.01
Blue Liquitint dye				0.03
HCl	0.885	0.885	0.885	0.84
CaCl₂	0.4	0.8	1.2	0.4
H₂O	balance to 100	balance to 100	balance to 100	balance to 100
Physical characteristics				
pH	2.6	2.8	—	2.5
Softening efficacy (EQ)	>40 EQ	>40 EQ		40 EQ
Initial viscosity (cP) spindle 3, 60 rpm	176	384	Thick Gel	128
Week 1	192	496	—	175 (week 17)

phase at 70°C with the water phase at 70°C containing hydrochloric acid. The $CaCl_2$ solution was added at ~60 to 70°C and the mixing rate was reduced to 300 to 500 rpm, and the system was cooled down to room temperature with an ice/water bath.

5. Preparation of AA/DEQ(H) Emulsions

Varisoft 510 AA(H), Varisoft 511 AA(S), and DEQ(H) were each melted, mixed together (composition 27, Table 5) with stirring, and maintained at 70°C. Perfume (Douscent 653 from IFF) was added to the molten mixture just prior to emulsification with the aqueous phase. Separately HCl solution was added to heated (~80°C) deionized water. The mixture of molten softening active compounds was added to the acidified water phase at 70°C with stirring using a 4-pitched blade impeller (1000 g batch, propeller stirrer, diameter of stirrer 100 mm; diameter of shaft 8 mm, length 500 mm). During addition of the molten mixture the stirring speed was increased from 300 to 700 rpm as the emulsion thickened. Calcium chloride (0.4%, as 20% aqueous solution) was added to break the gelation. The hot mixture was stirred for an additional 10 min at 350 rpm and the emulsion was allowed to cool to 30°C at which time colorant (0.03%, Milliken Liquitint Royal Blue) and preservative were added.

IV. RESULTS AND DISCUSSION

A. Difatty Amidoamine (AA) Dispersions

The AA molecule is the amine precursor to the traditional softener difatty diamido ammonium methyl sulfate. The AA(H) was selected considering its good environmental test results, favorable acute toxicity data, and performance (1:1 emulsions) [12]. The 1:1 regular strength system at 5% level in the presence of Armeen DMHTD (hydrogenated tallow dimethyl amine) as an emulsifier delivers better softening than a 5% dispersion of DHTDMAC. Keeping the ratio of AA(H) to Armeen DMHTD of 6.7, several emulsions were prepared. It was found that above 11% level, AA(H) dispersions were thick, nonpourable, nondispersible gels (Fig. 3), even in the presence of the electrolytes such as NaCl and CaCl$_2$, which are traditionally used as viscosity-controlling agents for softener emulsions [13,14].

It is presumed that the limitation (no more than 11% by weight before gelation) of the concentratability of the AA(H) emulsions is the result of the crystallinity of the fatty tertiary amine, combined with the formation of the multilayered vesicle structures. This may be analogous with the dihydrogented tallow dimethyl ammonium chloride dispersions or the phospholipid dispersions where the dispersed phase consists of multiwalled vesicles and the liquid crystalline phase is partially preserved in aqueous dispersions [14].

Low-viscosity concentrates (~10% to 20%) of fabric softeners such as imidazoline derivatives and other dialkyl ammonium compounds have been prepared by making use of high-pressure homogenization [15]; fatty alcohols, aliphatic fatty acids, and fatty acid methyl esters [16]; paraffin oils and paraffin waxes [17]; ionogenic and nonionogenic emulsifiers (such as ethoxylated nonylphenol

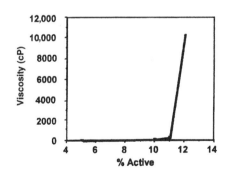

5%   Varisoft-510
0.75% Armeen DMHTD
0.36% HCl
2%   2-Propanol
0.02% Defoamer
91.88% Deionized Water

FIG. 3 Composition of 1:1 emulsion and the Brookfield viscosities of AA(H), Varisoft-510 dispersions (5% to 12% by weight) in the presence of hydrogenated tallow dimethyl amine (Armeen DMHTD).

and oleylalcohol) [18]; and water-miscible organic solvents such as hexylene glycol [19]. A few other examples of softener concentrates include the use of water-soluble polymers [20,21], alkoxylated alcohols [22,23], organic solvents [24,25], and fatty groups (in diester quaternary ammonium compounds) having controlled levels of unsaturation [26].

In order to prepare the low-viscosity concentrates of AA(H), two complementary basic approaches were made:

1. Reduction in the crystallinity of AA(H) by low-melting-point softening molecules such as amidoamine AA(S), unsaturated imidazolinium quat, and triethanolamine diester quat (DEQ).
2. Use of electrolytes as viscosity-controlling agents.

With the use of AA(S) and calcium chloride, the critical concentration (the concentration above which the viscosity increases to unacceptable values, i.e., >700 cP at room temperature) of AA(H) dispersions was shifted dramatically from 11% to about 25% (Fig. 4).

AA(S) is an analogue of AA(H), with soft-tallow moieties. In soft-tallow group, there is about 45% of unsaturated alkyl chains [mainly, the oleyl moiety $CH_3(CH_2)_7CH=CH(CH_2)_7COR$] and 55% of saturated alkyl chains ($C_{14}-C_{18}$). Conversely, in hydrogenated tallow (also sometimes referred to as "hard tallow" or "H-tallow"), all of the unsaturated chains are converted to saturated chains. Therefore the AA(H) molecule contains almost 100% of saturated alkyl chains. The presence of soft-tallow moiety in amido amine molecule changes the melting point of the amine from ~66°C (AA(H)) to ~41°C (AA(S)).

Blending of AA(H) with AA(S) presumably lowers the crystallinity of AA(H)

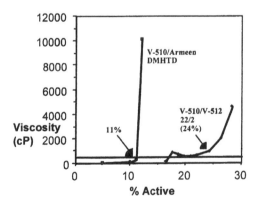

FIG. 4   Shift of the critical concentration of AA(H), Varisoft-510 emulsions from 11% to 24% by AA(S), Varisoft-512, and CaCl$_2$.

and thus provides a channel to go to relatively more concentrated dispersions (~25% by weight; Fig. 4) in the presence of $CaCl_2$.

Figure 5 shows the softening efficacy of various AA(H)/AA(S) emulsions. The softening deteriorates as one goes to higher concentration. While the 4% dispersion shows a product activity of 5 EQ, a 24% dispersion exhibits a softening of only 14 EQ. Probably this reduced softening is related to the higher average particle size and the remains of undispersed amine. For a range of 4% to 17% systems the average particle size is in the range of 1 to 2 μm. On the other hand, the particle size grows to 17 to 33 μm for 18% to 22% dispersions. Relatively smaller particles can uniformly deposit on the fabric, covering a higher surface area and causing a superior softening effect. Freeze-fracture electron micrographs of the composition 11 show some solid particles which are trapped inside the multilamellar vesicles. These solid particles are probably of unprotonated fatty amine. This probably means that fully dispersing the AA(H) may in turn enhance the softening of the emulsions.

B. Difatty Amidoamine/Difatty Amido Imidazolinium Methyl Sulfate Dispersions, AA/AIm

Another choice to reduce the crystallinity of AA(H) was with difatty amido imidazolinium compound (AIm; Fig. 6). This compound was chosen since it is liquid (90% active in isopropanol) at room temperature and because it may increase the dispersibility of AA (H) by reducing its crystallinity.

Evidence of the reduction in the crystallinity of AA(H) [or Varisoft-510], by AIm comes from differential scanning calorimetric analysis (DSC) of AA(H)

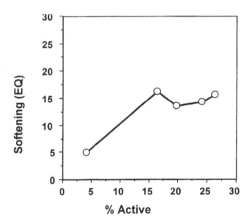

FIG. 5 The softening efficacy of various (AA(H)/AA(S)) emulsions (Table 2, compositions 5–12; the 4% active system consists of 3.5% AA(H), 0.5% AA(S), and 0.25% HCl).

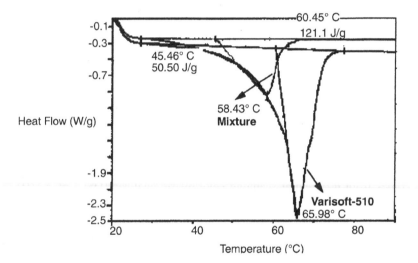

CH₃

TOCHNH₂CH₂C

AIm
T = Soft Tallow

FIG. 6   Imidazoline structure.

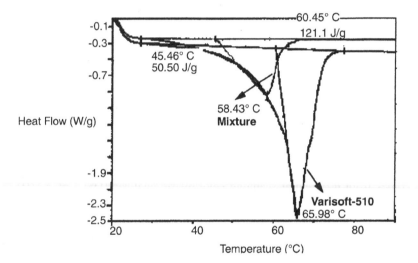

FIG. 7   DSC thermogram of Varisoft-510 AA(H), and 1:1 mixture of AA(H)/AIm.

and a mixture of AA(H) and AIm (1:1) (Fig. 7). DSC is a technique that measures the fundamental quantity of energy change during cooling or heating on a small quantity (5 to 10 mg) of sample. Thermogram of AA(H) shows a melting temperature of 65.98°C (DH = 121.1 Jg⁻¹). In contrast, a 1:1 mixture of AA(H) and AIm exhibits a melting temperature of only 58.43°C (DH = 50.50 Jg⁻¹). Therefore AIm serves as a dissolving matrix for the highly crystalline AA(H).

The characteristics of 1:1 mixtures of AA(H) and AIm are listed in Table 3. The unique combination of fatty amidoamine AA(H) with the imidazolinium quat exhibits a synergistic softening of 35 EQ at 25% level, an average particle size of about 1 μm, and a 1-week viscosity of only 360 cP at room temperature

(16). In contrast, dispersions based on solely AIm (17) or a combination system of AA(H) and AA(S) (15) deliver relatively poor softening efficacy. For example, AIm emulsion at 20% level shows only a softening of < 8 EQ (17). At first it was thought that the low level of softening efficacy of AIm is related to the relatively larger average particle size, 12 μm. However, no improvement was found in the softening for a homogenized sample either where the pH was adjusted to 2 (sample 17 has a pH of 5). So inherently AIm exhibits poorer softening than DHTDMAC, and this could be related to the lower hydrophobicity of the molecule due to unsaturation in the alkyl moieties. The commercial softening products generally do not use all soft-tallow-based softeners because of low softening.

The combination system of fatty amidoamine AA(H) and imidazolinium quat (AIm) meets the need of high-efficacy and low-viscosity fabric conditioner concentrate. The high efficacy and low viscosity of AA(H) in the presence of unsaturated softening molecules are ascribed to the high dispersibility of AA(H), the decrease in the melting point, and looser packing of the softening molecules in solution. A U.S. patent has been granted on AA/AIm work [27].

C. Difatty Amidoamine/Quaternized Triethanolamine Diester Quaternary Methyl Sulfate Emulsions AA/DEQ

The key learning of the AA(H)/AIm work was that *the introduction of unsaturation with softening molecules of dissimilar structure to the fatty amidoamine increases the efficacy and the concentratability of the softener emulsion.* This learning became the basis for the selection of the next combination system based on environmentally safe fatty amidoamine, AA, and triethanolamine diester quaternary methyl sulfate (DEQ and DEQ-H) [11,12,28,29]. It was possible to develop concentrates with as high as 36% actives (Fig. 8). The emulsions of fatty amidoamine and DEQ were of low viscosity (<600 cP) and exhibited acceptable viscosity stability with time (Tables 4, 5).

These concentrates delivered excellent softening efficacy as compared to DT-DMAC and AA emulsions. The 34% emulsions 22 (AA(S)/DEQ) and 24 (AA(H)/DEQ) exhibited softening efficacies of 35 and >>40 EQ, respectively.

To understand the role of unsaturation, a systematic study was carried out in which the amount of soft-tallow amidoamine AA(S) was varied from 0% to 100% in the amidoamine fraction (AA(H) + AA(S)) and monitored the impact on the physical properties of the AA/DEQ(H) emulsions (Fig. 9).

Figure 9 shows the effect of concentration of AA(S) on the viscosity of the AA/DEQ(H) emulsions. As can be seen, as the soft-tallow amidoamine, AA(S), increases, the viscosity decreases. These data correspond to samples stored at 4°C for 3 weeks (viscosity measurement at room temperature). The data clearly point out that in order to avoid gelation, the amount of soft-tallow amidoamine in the amidoamine fraction must be >45%.

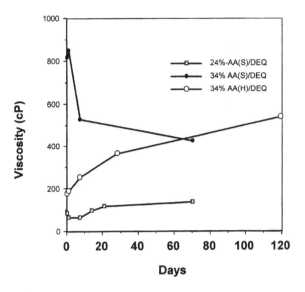

FIG. 8 Viscosity variation of AA/DEQ emulsions with time (Tables 4, 5).

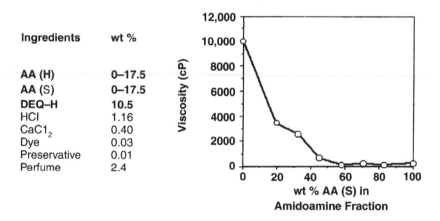

FIG. 9 Effect of concentration of AA(S) on the viscosity of the AA/DEQ(H) emulsions.

One theory for the decrease in viscosity is the reduction in the crystallinity (loosely packed structure as opposed to well-packed structure) of AA/DEQ(H) emulsions. The evidence for the reduction in the crystallinity of amidoamine emulsions comes from differential scanning calorimeteric analysis of emulsions (Fig. 10).

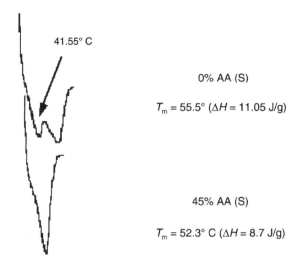

41.55° C

0% AA (S)

T_m = 55.5° (ΔH = 11.05 J/g)

45% AA (S)

T_m = 52.3° C (ΔH = 8.7 J/g)

FIG. 10 Differential scanning calorimetry heating thermograms of compositions (Fig. 9). Heating rate = 10°C/min.

The DSC data for the aqueous compositions all reveal multiple endothermic peaks. For simplicity, only peaks due to the chain melting are shown. The AA(H)/DEQ(H) emulsion with 0% soft-tallow amidoamine shows two distinct endothermic transitions at 41.6°C and 55.5°C. These peaks are probably due to the chain melting of DEQ(H) and AA(H), respectively. As the amount of soft-tallow amidoamine increases, these peaks merge to a single transition. This transition further shifts to low values; for example, the chain melting occurs at 47.5°C for an emulsion with 100% soft-tallow amidoamine, AA(S). The overall calorimetric enthalpy changes are also lowered as one increases the amount of soft-tallow amidoamine in the emulsions.

The decrease in the melting point and the ?H (heat of reaction or transition, or enthalpy of reaction) both suggest a lower degree of hindrance for the phase transition step; this in turn may relate to break up of the structure of the lamellar gel-phase.

In order to see the effect of unsaturation on the flow properties of the emulsions, a number of rheological techniques were employed. Figure 11 shows the flow curves of emulsions with various levels of AA(S). As the amount of AA(S) increases, the yield stress of the system decreases and the emulsion moves toward Newtonian (waterlike) behavior.

Yield stress refers to the minimum stress (force) that is required to initiate flow. Below this stress, the system behaves like an elastic solid. Relatively higher value of yield stress for the emulsion with 0% soft-tallow amidoamine is

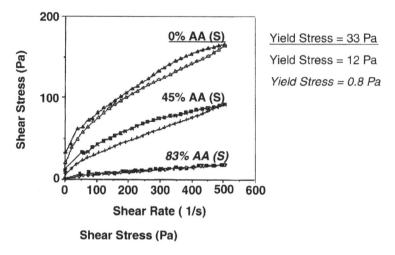

FIG. 11   Flow curves of 28% emulsions with various levels of soft-tallow amidoamine AA(S) in the amidoamine fraction.

indicative of stronger interactions between the particles. The incorporation of AA(S) certainly weakens the intermolecular interactions, leading to weakly or poorly structured systems.

The oscillatory measurements were performed at room temperature within the linear viscoelastic region to probe the "at rest" structure of the above emulsions. In general, within the linear viscoelastic region, both G' and G'' are independent of strain and the moduli are only functions of temperature and frequency. The G' (storage modulus) is a measure of the energy stored and retrieved when a strain is applied to the composition, while the G'' (viscous or the loss modulus) is a measure to the amount of energy dissipated as heat when strain is applied. In other words, the storage moduli, G', are a measure of the elasticity of the systems. The loss moduli, G'', are representative of the viscous behavior of the compositions. Figures 12 and 13 show the dependence of storage and loss moduli on the oscillation frequency (frequency sweep experiment, at fixed strain of 0.02) and torque (strain or torque sweep experiment, at fixed angular frequency of 6.283 rad/s) for samples (0% AA(S)) and (58% AA(S)). These figures point out that the storage moduli are higher than the loss moduli at all frequencies and torque values for an emulsion with 0% AA(S). This means that the concentrated dispersion, having 0% AA(S), is an example of a viscoelastic system. In contrast to this system, emulsions where the AA(S) content was in excess of 45% responded with more viscous than elastic response.

So the dynamic oscillatory measurements further verify the gradual loss of the structure with increased amount of unsaturated amidoamine. Figure 14

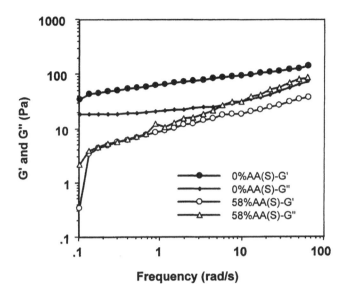

FIG. 12 Dependence of G′ and G″ on the angular frequency (rads⁻¹) for samples (0% AA(S)) and (58% AA(S)) (Fig. 9).

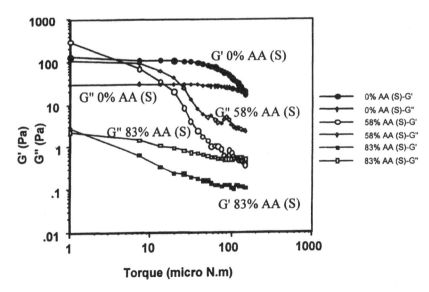

FIG. 13 Plots of G′ and G″ against torque for samples (0% AA(S)) and (58% AA(S) and 83% AA(S)) (Fig. 9).

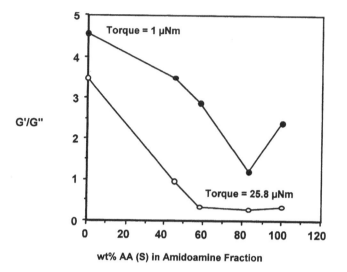

FIG. 14 Variation of the ratio of G'/G" with the concentration of soft-tallow ami-doamine, AA(S) in the amidoamine fraction (Fig. 9).

shows the effect of the concentration of soft-tallow amidoamine in the ami-doamine fraction on the ratio of G'/G" at a fixed torque (1 μNm and 25.8 μNm; torque sweep experiment at a frequency of 6.282 rads^{-1}). This ratio is a good tool to differentiate between samples that have more structuring than those that have less structuring or more thickening. The ratio decreases with the increase in the concentration of the AA(S) content in the amidoamine fraction. When G'/G" is >1, structuring predominates. Thickening or nonstructuring is favored for systems with G'/G" < 1.

For oil-in-water emulsions, more structuring is induced probably due to more lamellar phase volume where a higher amount of water is trapped inside the vesicles. Higher phase volume would result in relatively higher viscosity. Unsaturation seems to lower the lamellar phase volume; thus in turn it breaks the structure.

In order to find out the structural differences at microscopic level, some of the emulsions were studied by electron microscopy. In Figure 15 electron micrographs are shown for samples [0% AA(S), 58% AA(S), and 83% AA(S); Fig. 9]. With increase in the content of AA(S), the continuous electrolyte phase becomes visible, indicating a decrease in the dispersed phase volume. All micrographs show some multilayered structures (vesicles) at higher magnification. Densely packed domains of lamellar structures are probably responsible for the relatively strong elastic nature of the composition with 0% AA(S).

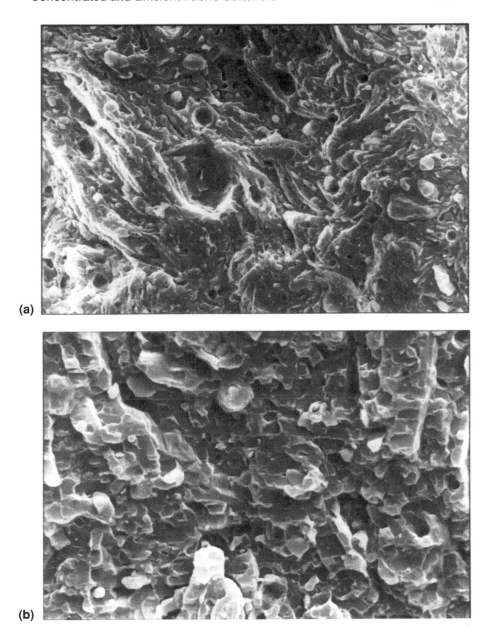

(a)

(b)

FIG. 15 Low-temperature scanning electron micrographs of 28% emulsions (a): (0% AA (S)), (b): (58% AA(S)), and (c): (83% AA(S)) (Fig. 9).

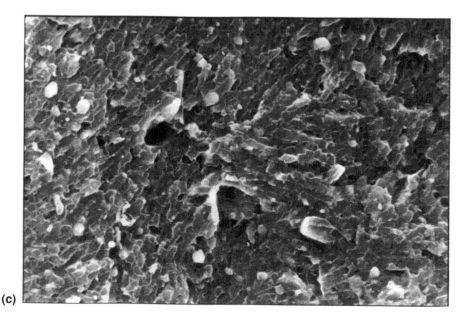

(c)

FIG. 15 Continued

Now that the experiments have determined impact of unsaturation on the viscosity and structure of the concentrates, it was important to find out how the softening is affected with the concentration of the AA(S) (Fig. 16). As expected, the softening decreases with the increase in the concentration of the unsaturated amine. In order to maintain low viscosity and have product activity of at least 40 EQ, it is preferred to limit the amount of soft-tallow amidoamine (in the amidoamine content) to no more than about 70% and no less than about 50%. The eight-times-diluted dispersion (3.5%) (composition 27) exhibits a softening efficacy of 5 EQ and an average particle size of 0.32 μm as determined by Brookhaven laser light scattering instrument at 90° angle (only 0.3% particles of size >2.5 μm as determined by Hiac/Royco series 4300 instrument). The structure of AA/DEQ dispersion at the microscopic level is totally different than the classical quat (DTDMAC) emulsion (5 wt%) (Fig. 17). The quat (DTDMAC) dispersions consist of large spheroidal particles (of size in the range of 5 to 20 μm), called vesicles. The internal structure of vesicles resemble that of an onion. As compared to DTDMAC the AA/DEQ dispersion consists of relatively small particles and lacks such onion-shaped multilayered big particles. The better softening (equal weight basis) for AA/DEQ system is ascribed to the presence of relatively small particles, which probably uniformly deposit on the fabric surface. Highly concentrated low-viscosity formulation is achieved in case of AA/DEQ due to

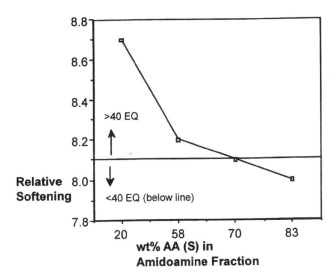

FIG. 16 Effect of unsaturation on the softening of compositions (Fig. 9) (8-times-diluted samples, softening by DHTDMAC (5% dispersion) gives a value of 8.1).

the availability of more free water in the dispersion. In case of quat (DTDMAC) a lot of water is tied up within the layered structure. A U.S. patent has been granted on the work based on AA/DEQ [30].

V. CONCLUSIONS

Selected combination of AA/AIm and AA/DEQ has been identified as one of the chemistries of choice for preparing premium fabric conditioner ultra concentrates. The viscosity of a concentrated colloidal dispersion based on AA, bis-(alkyl amidoethyl)-2-polyethoxyamine and DEQ, bis-(alkylcarboxyethyl)-2-hydroxyethyl methyl ammonium methyl sulfate (triethanolamine diester quaternary methyl sulfate) can be reduced by the incorporation of unsaturation within the fatty groups of the molecules. In this study it was discovered that the degree of unsaturation influences the dynamic mechanical properties of concentrated cationic surfactant dispersions. Results obtained using differential scanning calorimetry, scanning electron microscopy, and rheometry show that the rheological behavior of dispersions is related to the breakup of the viscoelastic structure as a function of the degree of unsaturation within the alkyl groups of the cationic surfactants. The AA/AIm and AA/DEQ systems are found to be more efficient softeners than DTDMAC. This finding is related to and confirms other work on the advantages of combinating amine salts and quats [31,32].

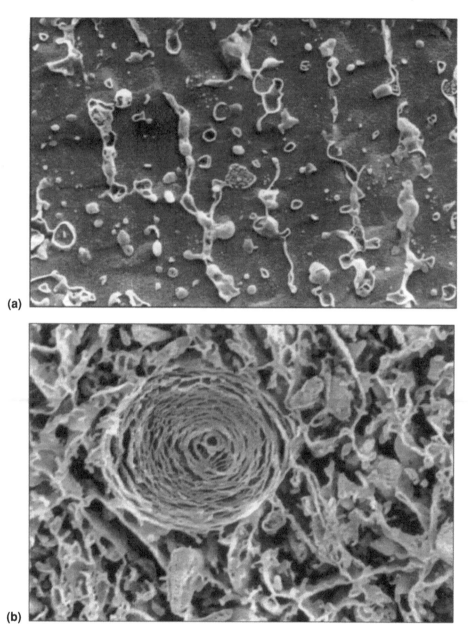

FIG. 17 Low-temperature scanning electron micrographs of 8-times-diluted sample of 28% dispersion (a) (composition 27) and 5% dispersion (b), of DHTDMAC.

ACKNOWLEDGMENTS

We express our sincere appreciation to Charles Schramm, Alain Jacques, and Robert Steltenkamp for helpful discussions; Christine Bielli, Francoise Deville, and Karla Tramutola for providing excellent technical assistance. We thank Suman Chopra and Rodman Heu for their help in obtaining differential scanning thermograms and electron micrographs, respectively.

REFERENCES

1. A Jacques, C Schramm. In: K-Y Lai, ed. Liquid Detergents. New York: Marcel Dekker, 1997, pp 433–462.
2. RP Adams. U.S. patent 4,741,842 to Colgate-Palmolive Company (1988).
3. AP Murphy. U.S. patent 4.239,659 to Procter & Gamble Company (1978).
4. TE Cook, EW Dolle. U.S. patent 4,493,773 to Procter & Gamble Company (1984).
5. R Lagerman, S Clancy, D Tanner, N Johnston, B Callian, F Friedli. J Am Oil Chem Soc 71:97, 1994.
6. GR Whalley. HAPPI February:55, 1995; and FE Friedli, RO Keys, PH Orawski. Present status of fabric softener actives. World Congress for ISF, Kuala Lumpur, 1997; and MI Levinson. Fabric softeners today and tomorrow. Soaps Detergents & Oleochemicals Conference, Ft. Lauderdale, FL, 1997.
7. J Waters et al. Tenside Surf Det 28(6):460, 1991; ES Baker. Preparation, properties, and performance of DEEDMAC, an environmentally friendly cationic softener. AOCS Convention, Atlanta, 1994.
8. RB McConnell. INFORM 81:76, 1994.
9. W Ruback. J Conite Esp Deterg 181–192, 1989.
10. H Hein. Tenside Detergents 18:243, 1981.
11. RO Keys. HAPPI March:95, 1995.
12. C Schramm, R Cesir. U.S. patent 5,476,598 to Colgate-Palmolive Company (1994).
13. RG Laughlin. In: DN Rubingh, PM Holland, eds. Cationic Surfactants. New York: Marcel Dekker, 1991, pp 449–465.
14. AD James, PH Ogden. J Am Oil Chem Soc 56:542, 1979.
15. JA Monson, WL Stewart, HF Gruhn. U.S. patent 3,954,634 to S.C. Johnson & Sons (1976).
16. M Verbruggen. EP patent 0013780 A1 to Procter & Gamble Company (1980).
17. PCE Goffinet. British patent 1 601 360 to Procter & Gamble Company (1981).
18. H Hein. Tenside Detergents 18:243, 1981.
19. NA MacGilp, AC McRitchie, BT Ingram, J Hampton. U.S. patent 4,454,049 to the Procter & Gamble Company (1984).
20. SQ Lin, LS Tsaur, LV Sales. U.S. patent 5,429,754 to Lever Brothers.
21. JG Pluyter, MG Eeckhout. EP patent application 0705900 A1 to Procter & Gamble Company (1996).
22. NJ Chang. U.S. patent 5,066,414 to Procter & Gamble Company (1991).
23. SR Ellis, GA Turner. AU-B-13125 patent to Unilever (1992).

24. DF Groot. WO (94) 17169 A1 patent application to Unilever (1994).

25. T Trinh et al. WO (97) 03170 patent application to Procter & Gamble Company (1997).

26. EH Wahl et al. WO (94) 20597 patent application to Procter & Gamble Company (1994).

27. A Farooq, A Mehreteab, G Broze, C Bielli. U.S. patent 5,468,398 to Colgate-Palmolive Company (1995).

28. L Contor, Y Lambremont, C Courard, P Rivas. U.S. patent 5133885 to Colgate-Palmolive Company (1992).

29. H Birkhan, H-J Köhle, J Weigand, W Wehner. U.S. patent 5,180,508 to Rewo Chemische Werke GmbH (1993).

30. A Farooq, R César, F Deville. U.S. patent 5501806 to Colgate-Palmolive Company (1996).

31. ME Burns. U.S. patent 4,439,335 to Procter & Gamble Company (1981).

32. T Trihn, EH Wahl, DM Swartley, RL Hemingway. U.S. patent 4,661,269 to Procter & Gamble Company (1985).

Index

T - #0044 - 111024 - C0 - 229/152/17 - PB - 9780367397586 - Gloss Lamination